WORKING WELL

Managing for Health and High Performance

Marjorie Blanchard, Ph.D.

Mark J. Tager, M.D.

Simon and Schuster

New York

SIMON AND SCHUSTER and colophon are registered trademarks
of Simon & Schuster, Inc.
Designed by Irving Perkins Associates
Manufactured in the United States of America
1 2 3 4 5 6 7 8 9 10

Library of Congress Cataloging in Publication Data
Blanchard, Marjorie, date.
Working well.

Bibliography: p.
1. Industrial hygiene—Management. 2. Job stress.
3. Work environment. I. Tager, Mark. II. Title.
HD7261.B58   1985   658.3'14   85-14644
ISBN: 0-671-54564-7

ACKNOWLEDGMENTS

Working Well could never have been created without the help of our colleagues and friends. Each contributed his or her unique abilities and leadership styles. Heartfelt thanks to our:

> *Critical Reviewers (S1)—Dr. Patricia Zigarmi at Blanchard Training and Development, Inc. (BTD), Drs. Barry Cohen and Richard Bellingham from Possibilities, Inc., Stephen Willard of Asian Business Consultants, Andrew Workhum from Westinghouse Electric Corporation, and Ahouva Steinhaus. They gave us high direction and plenty of task-specific feedback.

> *Coaches (S2)—Emmit McHenry of Allstate Insur-

ance, Ron Semone from the Veterans Administration, Norine and Kelsey Tyson at BTD, Molly McCauley of AT&T Communications, and Harry Safstrom of Dow Chemical. Their guidance and encouragement helped us through some of the rough times.

*Supporters (S3)—Our medical colleagues Drs. Joe King and Wayne Burton, Pam Strickfadden from Benjamin Franklin, Dr. Drea Zigarmi, Linda Hendricsen and Eleanor Terndrup from BTD. They proved a great sounding board for our ever-changing thoughts and ideas.

*Believers (S4)—Our families, who have always believed that no matter what we did, it would work out all right.

Finally, special thanks to two members of our "management team" who stuck with us through the whole development cycle:

*Our associate, Barbara Terman, whose help in research and editing was invaluable, whose patience and flexibility was just a bit short of heroic, and Fred Hills, our editor at Simon & Schuster, who bought a book based on a belief in its authors.

CONTENTS

9

Contents

Contents

FOREWORD

by *KENNETH BLANCHARD, Ph.D.*
Co-author, THE ONE MINUTE MANAGER

When Spencer Johnson and I wrote *The One Minute Manager*, we highlighted a major truth about business: PEOPLE WHO FEEL GOOD ABOUT THEMSELVES PRODUCE GOOD RESULTS. As a management consultant and trainer, I have known for a long time that managers have a powerful influence on the attitudes, feelings, and mental well-being of their people. Managers who appreciate and support their people's efforts are likely to have employees who feel good about themselves and their jobs, people who go to work with a skip in their step and a twinkle in their eye. In the language of the One Minute Manager, their managers "catch them doing something right."

On the other hand, when a manager is unapprecia-

15

tive, punishing, and accents the negative, employees are much more likely to dread work, perform well only when closely supervised, and live by the motto "Thank God it's Friday."

In talking to managers and workers in all areas of the country, I hear far more stories of people who are treated badly on the job than of those who are treated well. I also listen as companies tell me their "war stories" about the financial effects of poor health. Saddest of all, I see the stress and performance problems that result from people being poorly managed. I'm struck by the tremendous waste of human resources from these combined problems, knowing full well they could be turned around if managers knew what to do.

Working Well is a breakthrough book, because it examines these problems from the standpoint of increasing your people's potential. Mark Tager, an M.D. with recognized expertise in organizational health promotion, has long emphasized that maximum *wellness* means more than the absence of disease; it includes attitudes, feelings, and ability to perform. Marjorie Blanchard, a Ph.D. with extensive experience in management education and communications, knows from firsthand observation the effect that a manager can have on an employee's mental well-being and productivity. Combining the two viewpoints, they took a fresh look at the role of a manager in keeping employees healthy *and* productive. They began to ask, "Can a bad manager make people sick?" The answer, from workers across the country, is a resounding "Yes!"

That's where the term "working well" comes in.

Foreword

Through their combined research and observations, Margie and Mark have come up with a five-part formula that describes what every manager can do to encourage "working well" in his or her organization. Its ingredients are called PERKS, a set of practical guidelines that help managers to make a long-term commitment to cultivating and protecting human resources without sacrificing productivity.

I found PERKS a simple, important, easy-to-understand set of guidelines for more effective management. Managers who encourage *Participation* among employees, create a supportive *Environment*, provide *Recognition* for efforts, openly share *Knowledge* that affects their people, and continually try to match their leadership *Style* to the needs of employees will be able to transform potentially harmful stress into productive, creative energy.

All in all, I found *Working Well* a fascinating book that will challenge you, make you think, and—if you take heed of its suggestions—make you a much better manager of people and productivity.

How This Book Will Help You

"Our *people* are our most important resource." How many times in recent years have you heard variations on this theme? In advertising and internal communications, organizations emphasize the message that hardworking, committed employees are the key to the organization's success.

Yet when it comes to day-to-day management, many organizations misuse their "most important resource." Managers accomplish short-term objectives, but their people pay a high price; they become burned out, disillusioned, and ultimately less productive.

It doesn't have to be that way. When "human resources" are managed properly, everyone benefits. Employees are physically healthy, enthusiastic, hardworking, and productive. We have a word for this special combination of physical health, good attitudes, and increased productivity. We call it *wellness*, a term that originated in the health-care field to describe a positive state of physical and mental well-being that includes more than just the absence of illness. Wellness emphasizes self-responsibility for the whole range of factors that contribute to both physical and mental health. On the physical side, it encompasses fitness, nutritional awareness, and stress management—in fact, everything that contributes to a sound, energetic body.

But wellness doesn't stop at feeling physically good. It also includes the psychological traits that contribute to mental stamina—the internal values that contribute to *working well*. Employees at a high level of wellness are more creative and productive, communicate better, and have a generally positive outlook toward life.

The link between employee wellness and "working well" is what we call the health/productivity connection. In recent years, organizations have become increasingly concerned about the *negative* side of this equation, that is, the costs of *poor* health and *lowered* productivity. Businesses have good reason to be concerned: They are footing the health-insurance bill for 140 million Americans—a bill that is substantial and growing at an alarming rate:

- American companies pay $100 billion a year in total health-care costs, including insurance and preventive programs. This is twice what they paid five years ago.

- In the average Fortune 500 company, medical bills run about 24 percent of after-tax profits. Health-insurance premiums are increasing about 20 percent a year.

- Health-care costs are the second-largest expense (after payroll) in service businesses. They are the third-largest expense (after salaries and materials) in all industry.

- 500 million workdays are lost each year due to illness or disability.

- Alcoholics and smokers have twice as many sick days as other employees.

That's the *bad* news. But there is a positive side to the health/productivity connection, one that is the basis of this book. Briefly, the "good news" about the health/productivity connection can be expressed in a single sentence: *Healthy people produce better results*.

The *working well* formula describes what every manager can do to create wellness in his or her people. Its ingredients are known as "PERKS," a set of practical management guidelines which allow you to make a long-term commitment to cultivating and protecting human resources without sacrificing productivity.

MANAGING THE WHOLE PERSON

Over the last decade both authors have worked with organizations around the country installing programs to improve people's performance and well-being. Marge's company, Blanchard Training and Development, gives managers the skills they need to be more effective leaders; Mark has helped companies set up programs in fitness, stress, and other health-related areas. Although we come at the issue from different angles, both of us have had the opportunity to work with thousands of individuals on the path to wellness.

As we began to discuss our work together and to see the results of the other's labors, we noticed a synergism between our two efforts: People who were well managed were healthier, and healthier people produced better results.

The link between management principles and employee health shows up all around us. You instinctively know that people work better when they're healthy, and statistics from study after study back up this notion. Your people's good health isn't just a "nice thing" for them—it's a real productivity benefit for the company in terms of absenteeism, health-care costs, turnover, and morale.

On the other hand, you also know that people feel physically better when they're well managed. If you set clear-cut goals and rewards, give generous feedback and praise, and use the appropriate management style for a given situation, people will feel secure and in control. Unpredictability, crisis management, and lack of feedback, on the other hand, contribute to a number of stress-induced physical, mental, and emotional illnesses.

PERKS: THE TOOLS FOR INDIVIDUAL ACTION

In developing the **PERKS** system, we started with two questions that are central to our work: What makes people work hard? And what keeps them healthy? In an attempt to find common ingredients we have examined literature dealing with productivity, motivation, job satisfaction, and health promotion, and have drawn on our own experience. The result is **PERKS**, a set of health-promoting management techniques that help managers at every level bring out the best in their people.

PERKS describes a set of five skills and support tools that help your people "do better"—perform up to their capabilities, stay well, and maintain positive attitudes. They include:

> PARTICIPATION: *People do better when they are involved in the decisions that affect them.*
> ENVIRONMENT: *People do better when the environment gives them the opportunities and choices to perform well.*
> RECOGNITION: *People do better when they get feedback on their performance and recognition for their progress.*
> KNOWLEDGE: *People do better when they know where they're going, how to get there, and why they're going.*
> STYLE: *People do better when they are managed in a style appropriate for their level of skills and commitment.*

PERKS, of course, doesn't cover the entire range of management responsibility and behavior. Out of all the things a manager can do to improve productivity and make the worksite a better place to be, we have chosen five areas that represent the "80–20" rule. These are the 20 percent of activities that will provide 80 percent of the improvement in the workplace. There are any number of other techniques managers can or must use that will have an impact on productivity and the quality of work life, but PERKS represents the most effective way of improving the "people" side of the productivity equation.

To illustrate the PERKS principles throughout the book, we've used many anecdotes about managers and employees we've seen or heard of through our work. Both of us have conducted workshops and seminars in hundreds of worksites, intensively gathering material to illustrate the message of this book.

All of these illustrations are taken from experiences we have either witnessed or been told about. In cases where the manager's behavior has had a negative effect on employee health and performance, we've changed names and disguised company characteristics. We're not trying to condemn specific "bad" managers, just pointing out how their *behavior* can detract from employee health and productivity.

On the other hand, we believe that managers and companies that are actively involved in promoting wellness among their employees should be recognized. So when we describe a situation where employees are being managed in a way that maximizes both their health and productivity, you'll see the names of the managers and organizations responsible.

Chapters 3 to 7 explore Participation, Environment, Recognition, Knowledge, and Style in greater detail, sharing with you the practical guidelines that make PERKS effective for individual managers. In addition, the principles behind PERKS often apply to organized health-promotion efforts. Whenever appropriate, we've shown how managers can use PERKS to help establish or support these programs. For managers who wish more information, Part II, "Resources: Health-Promotion Programs," describes several successful programs and the key elements behind them.

BUILDING HEALTHIER PEOPLE MAKES EVERYONE WORK WELL

We hope that this book will challenge you to use your leadership position to bring out the best in your people. Wellness benefits *everyone*: You win by getting the results you want from your people, the company wins through increased productivity, and your people win by being managed for wellness.

Managing for Health Is Managing for Productivity

Chapter 1
Who Cares about Health?

- You've got an important presentation to give at noon. You were counting on your secretary to make last-minute corrections and prepare the information packets. She's just called in sick. Your personal productivity plummets as you neglect other projects to get your presentation ready.

- At last the new computer system is ready to become operational. Unfortunately, the start-up has been postponed indefinitely. The Director of Information Services had a heart attack, and the word is that he'll be out at least three months. You can practically see the orders that will be lost or delayed until the new system is running.

• A former employee believes that his anxiety and depression are the company's fault—the result of stress from his former job. He is now seeking to receive workers' compensation for what he believes is a job-related condition. You face the prospect of spending dozens of hours reviewing records, or tied up in meetings with the legal staff and personnel department. You do a "slow burn" as you think of the resources pulled away from other essential tasks to work on this case.

Mention the word "health" to most business people and what comes to mind? Usually thoughts of health-care benefits or annual checkups. In the course of an average day, few managers think about health. After all, their main concern is getting the work done. Their emphasis is on making schedules, creating new products, providing quality service—functions that everyone would agree are central to their jobs. In contrast, health seems like a dull subject, not relevant to the "real" business at hand.

Yet, as the cases above show, health can have a direct and disastrous impact on getting the results you want. When illness or disability strikes, it can become a manager's nightmare.

THE HEALTH/PRODUCTIVITY CONNECTION

The situations above illustrate anecdotally what everyone knows: Health problems can have a negative effect on productivity. But let's take a closer look at the health/productivity connection, to better understand its origins and how widespread its problems are.

> Absenteeism, premature death and disability, and lack of stamina and endurance have their roots in poor lifestyle habits.

These "poor habits" are what doctors and public health officials call *risk factors*. On the simplest level, risk factors are those elements that tip the odds away from wellness toward disease and premature death. While some risk factors (such as heredity and age) are not under our control, the most important factors are *behavioral*, relating to lifestyle choices in areas like exercise, nutrition, stress management, and safety.

Extensive medical research over the last two decades has examined the risk factors related to the most widespread and dangerous health problems—heart and lung disease, cancer, cirrhosis of the liver, diabetes, and accidents. In the figure below, we've applied some national statistics to a random group of 1,000 employees, to give you a better idea of the magnitude of the problem.

RISK FACTORS PER 1,000 EMPLOYEES

200–250 are overfat	300 are prone to low-back injury	1 out of 12 women will develop breast cancer

290 smoke cigarettes	**FOR EVERY 1,000**	100–150 have alcohol or drug problems
860 do not use seat belts	160–250 have hypertension—and half of these don't know it	500 are underexercised[1]

These risk factors can be thought of as time bombs ticking away among your people, which may explode into injury, illness, disability, or death. While there's no way of telling exactly how many of these time bombs will actually explode, it is possible to translate these risks into financial costs. The box below provides an estimate of the high costs of unhealthy lifestyles.

The High Costs of Lifestyle Disease

Diseases of the heart and blood vessels cost North American businesses about 29 million workdays per year. Losses due to health expenditures, disability payments, replacement and retraining costs have been estimated in the range of 39–52 billion dollars per year.

Employees who smoke cigarettes spend 81 million more days in bed and have double the occupational accident rate of non-smokers. Statistics indicate that businesses may be losing anywhere from $400 to $600 for each smoker they employ.

Back problems account for 27 percent of workers' compensation claims, amounting to $250 million in claims paid. The overwhelming majority of these problems are associated with poor physical conditioning.

Excess weight causes a rise in the risk of premature death—and the more overweight an individual, the greater the risk. People who are 20–30 percent overweight have a 20–40 percent greater risk than those at ideal body weight; those 50–60 percent overweight are 150–250 percent more likely to die prematurely.

Illegal drug use costs industry $26 billion a year. In addition to poor performance and increased absenteeism, the "typical" recreational drug user files five times more workers' compensation claims than non-users, uses three times more medical benefits, and is 3.6 times more likely to have an accident on the job. [2]

What's more, adding risk factors doesn't just add to the risk of disease—it multiplies. A person who smokes runs a two to two and a half times greater risk of developing cardiovascular disease than a nonsmoker. But a person who smokes and has high blood pressure may run up to ten times the chance of heart disease.[3]

Multiple risk factors also multiply health-care costs. According to a study conducted by the University of Michigan at Ford Motor Company manufacturing plants, the average annual health-care costs per employee multiplied with additional risk factors. The study compared two factors: high blood pressure and smoking. Health-care costs increased 45 to 55 percent when one risk condition was present—but by more than 115 percent for those with both conditions:

Cost per Year Risk Factors

$ 564	Normotensive (normal blood pressure) non-smokers
$ 818	Normotensive smokers
$ 870	Hypertensive (high blood pressure) non-smokers
$1,215	Hypertensive smokers [4]

Another risk factor that has commanded much attention over the last decade has been stress. In 1983, 11,600 employees filed workers' compensation suits for "work-induced" stress-related illness. In addition to mental symptoms of stress, the physical effects are far-ranging and serious—and potentially expensive. *Training Aids Digest* reports:

> ...in the State of Connecticut we have become aware of a sharp rise in the number of police officers who are being compensated for hypertension claimed to be the result of or caused by "on the job" stress. In most cases, municipal police departments must pay these claimants 50 percent of their salaries in compensation. Obviously, these accumulated expenses form a significant financial burden, resulting in very high insurance costs. [5]

These statistics lend objective proof to what every manager knows: poor health ultimately translates into problems with performance and productivity. Most of us don't need the statistics to appreciate the effects of poor health. We simply need to ask the question, How much do you get done when you don't feel well?

FROM PROBLEMS TO POSSIBILITIES

The link between health and the ability to perform on the job is instinctual. Top performers have always recognized that this connection is crucial to getting results. The positive side of the health/productivity connection was summed up for us by two leading executives:

William Smithburg, CEO of the Quaker Oats Company and an avid handball player for twenty-five years: "Staying in shape is like giving yourself added physical and mental endurance. You can handle the stress and strain of an active career better if you are in shape. It's both attitudinal and biological. I can't believe that a person who is in very poor shape is going to be as alert mentally as someone who isn't." [6]

Warren Batts, President of Dart & Kraft, Inc.: "Being fit helps a person in terms of stamina. I think that during the periods of time when I have gone two or three months without very much exercise there has been a noticeable difference in my energy level and probably an increase in my irritability factor. The key ingredients in this type of job are to think before you leap, have a long enough fuse to think things through, try not to go off emotionally when something doesn't go well. Fitness has a lot to do with the ability to stay in a sort of overview mode as opposed to getting too emotional about the ups and downs." [7]

Other top executives have taken their personal commitment to health a step further. In 1981 William M. Kizer, Sr., Chairman of the Central States of Omaha Insurance Company, mobilized other CEO's of the area to establish the Wellness Council of the Midlands, an organization devoted to improving employee health and curtailing rising health-care expenditures. Central States is just one of many companies throughout the country that have developed comprehensive efforts known as *health-promotion* or *wellness programs*.

WHAT ARE WELLNESS PROGRAMS?

They are a means of fighting back at rising health-care costs and at the underlying lifestyle problems that cause poor health. Health-promotion programs are organized company-wide efforts that provide opportunities for employees to increase their knowledge of wellness and acquire the skills and support to translate this knowledge into lasting behavior change.

These programs are often integrated into health-care cost-containment efforts such as restructuring of benefits, reviewing medical claims, selecting alternative health-care providers, and educating employees to be wiser health-care consumers. Health care costs the American consumer approximately $1 billion a day, with business and industry picking up about 23 percent of this tab. Every company is feeling the effects of this financial drain; few are now willing to passively accept the added fiscal burden. Health-promotion programs are an important component of an overall corporate health-management strategy.

What better place to provide health-promotion programs than at work? After all, most people spend approximately 40 percent of their waking time on the job. The attitudes and habits that are reinforced at work usually spill over into their personal lives as well. Besides, the worksite has several attributes that make it a natural place to center health-promotion efforts. The company has the resources to coordinate the program. The worksite is a convenient location for screening and monitoring results. The company is in a position to bring together both large numbers of employees and the expertise of various community resources for the delivery of education in the most cost-effective way.

Within this broad definition, wellness programs vary widely in their scope. The following chart shows part of the array of skills that may be taught and supported through programs at the worksite.

Health-Promotion Skills

FITNESS

Developing a balanced program
(muscle strength, flexibility and
cardiovascular endurance)
Getting started/staying started
Identifying target heart rate

Skills in recreational sports
Avoiding injury and recognizing
signs of overuse
Maintaining proper posture

NUTRITION AND WEIGHT LOSS

Making better nutritional choices
Reading food labels
Assessing the amounts of fat, sugar,
and salt in the diet
Learning healthy food preparation
techniques

Cooking low-fat meals
Making wise nutritional choices
when traveling
Becoming a better food buyer
Eating nutritiously on a low bud-
get

WEIGHT LOSS INCLUDES THE ABOVE PLUS:

Identifying calories in foods   Altering body image
Stimulus-response   Increasing caloric expenditure

STRESS MANAGEMENT

Identifying stressors   Stress relaxation techniques
Time management   Rational thinking
Self-dialogue   Assertiveness training
Communication skills   Support building
Imaging

Health-Promotion Skills (*continued*)

SAFETY

Lifting procedures
Strength and endurance exercises
Cardiopulmonary resuscitation (CPR)

Defensive driving
Attention control
First aid
Self-defense

MEDICAL SELF-CARE

Selecting a physician
Buying generic drugs
Treating minor conditions
Dental care, foot care, skin care
Breast self-exam
Home blood pressure testing

Communicating with health-care providers
Recognizing serious warning signs
Testicular self-exam
Wise use of health-care benefits

MANAGERS: SUPPORTERS OR SABOTEURS?

In the course of our consulting efforts we are often asked, Do these programs work? Are they effective? Our answer is an unequivocal yes. Although worksite health-promotion programs are a relatively new (within the last ten years) phenomenon, there are now enough well-controlled studies to document a number of positive benefits for both the individual and the organization, many of which translate into direct cost savings. Among the results are:

- decreased risk of dying from cardiovascular disease
- increased stamina and endurance
- improved coping skills in dealing with tension
- positive attitudes toward work
- greater commitment to the organization
- successful rehabilitation for substance abuse
- early detection and treatment of cancer
- better safety-related decisions

Ongoing studies are continuing to substantiate these findings; and for the manager interested in these statistics, we've included a number of "success stories" in Part II.

Despite the growing body of evidence that these programs benefit all levels of an organization, we have seen in our work that not all managers support such efforts. Or they may support health promotion in theory but find that it conflicts with other more immediate goals. Consider the way middle managers at a large telecommunications company unwittingly sabotaged a well-meaning health promotion attempt:

Top management had decided to implement a stress management program for employees at all levels. The CEO was behind the program 100 percent, and grassroots employee reaction was just as enthusiastic: 660 people signed up for the program. But when the seminar was given, only 120 people attended.

What went wrong? When the Human Resources Director began to investigate, he found that middle management and supervisors took an entirely different view of the program than the top managers and the rank-and-file employees. In theory, the supervisors thought a stress management seminar was a "good idea." But when it meant taking employees away from their work for several hours, they hit the roof.

These supervisors weren't "bad guys"; they were just living up to the letter of their job descriptions. They knew that their performance evaluations would hinge on the productivity of their departments, not on their support for health-promotion programs.

In this case top management had the right idea, but they didn't recognize the vital link between middle-management support and the success of the health-promotion program. They failed to communicate to the supervisors that health promotion was indeed a top priority.

That's where you, the manager, come in. The success or failure of a company-sponsored health-promotion effort often hinges on the *attitude* of employees toward the program. Are they skeptical and uninterested, or enthusiastic and dedicated to improving their own health? You are the vital link between an important organizational concern—building healthier people—and your people's involvement in these efforts.

One of the difficulties with introducing a wellness program to managers is that it often forces them to confront their own unhealthy habits. Depending upon how the program is introduced, this confrontation can ignite resistance, mainly because few of us want to be told that we are "wrong." On the one hand, we all know that living a healthy lifestyle is important; on the other, most of us have at least some unhealthy habits. How can a health program succeed if the most influential members of an organization, its managers, smoke cigarettes, are overweight, and don't get enough exercise? The answer is that *at the worksite health is more than a personal matter, it is an organizational objective and a management issue.*

In many organizations, long-range planning for employee health/productivity is seldom or never mentioned. Instead, short-range thinking and improving this quarter's bottom line are rewarded. Health-promotion programs are not a quick fix; they are a longer-term investment in your human resources.

There is a role for every manager in building a healthier organization. If you are in a position to institute policies and develop programs, or if your organization already has a health-promotion program in place, you can support these efforts. Regardless of whether you participate in formal activities, you can encourage others to do so, you can monitor the progress of your people, or join with other managers to establish programs.

Working Well includes suggestions for establishing the policies and guidelines to make certain these programs succeed. We encourage managers at all levels of an organization to get behind health-promotion efforts.

THE MISSING LINK

But there's a missing link in this picture, something that we have noticed in observing health-promotion programs around the country. Many of these efforts are divorced from the day-to-day realities most employees face on the job. Why? Because they fail to address one of the most powerful effects on health: the way people are managed.

> The manager/subordinate relationship is a major determinant of health

Studies around the country substantiate what every worker knows: A healthy boss/subordinate relationship is a major contributor to mental well-being and ultimately physical health. Companies which encourage employees to work off their stress in a gym or learn relaxation techniques to deal with frustrations are only providing one component of health management. These are valuable efforts, but they are often Band-Aid approaches to a much deeper health problem: *boss-induced illness*. As you'll see in the chapters ahead, there is no substitute for a healthy boss/subordinate relationship. Fortunately, every manager can learn to manage for wellness.

Chapter 2:
Are You Managing for Wellness?

*YOU AS A MANAGER EXERT A POWERFUL
INFLUENCE ON YOUR PEOPLE*

As a manager, you are important. You have more influence over your people than you may realize. They look to you for direction and guidance, acceptance and praise, feedback and reassurance. At the end of a day how employees feel, what they think, and what they have accomplished often depend on the type and quality of the interactions they've had with you.

In case after case, we've seen firsthand how a manager's influence can affect an employee's performance and attitude. To see this for yourself, watch a group of employees when their manager walks into the room. Does conversation immediately stop as everyone hurriedly tries to look busy? Do several people approach the manager saying, "Boss, I have a problem?" Do they greet the manager in a friendly, relaxed manner? All of these reactions say a lot about the manager/subordinate relationship.

Some managers want their people to be afraid of them. Others want employees to think of them as friends. Some control and direct every move; others give almost no guidance at all and leave people to fend for themselves. Any of these styles can be appropriate in the proper circumstances; the secret is knowing when and how to use each style.

But even those who understand the power of managerial influence often fail to see another vital connection: *How people are managed can directly affect their health.*

Many managers don't like to hear this. Those who are already overburdened with responsibilities will groan, "Oh, no. It's hard enough to get my job done as it is—now I have to keep my people healthy, too!"

But keeping people healthy doesn't have to be an extra burden. *Employee health and productivity are two sides of the same coin; the same techniques that will help boost productivity will also set the stage for better health.*

CAN A "BAD BOSS" MAKE PEOPLE SICK?

We have documented the positive and negative effects of a manager's influence in a number of ways. We've examined workers' compensation cases in which em-

45

ployees have brought suit for stress-related illness brought about by management practices.[1] We've studied the literature on how job stress can cause illness, where actual worksite situations were shown to have negative effects on people's health. We've investigated animal studies, which parallel many worksite situations but have the advantage of controlled conditions and measurable effects.

But perhaps our most interesting research has been simply asking people as we travel across the country, "Can a 'bad boss' make people sick?"

We use this question in workshop exercises where we ask people to think of the worst and the best managers they've ever had. How did the manager treat them? And how did they feel as a result? In performing this exercise, we pointed out that by "bad" boss we didn't mean a bad person but someone whose *actions* caused unnecessary tension or other negative feelings. By the same token, no one can "make" someone else sick, but the cumulative effect of daily stress from a poor manager/subordinate relationship can wear down energy and enthusiasm to the point where the employee actually becomes ill.

Can a "bad" relationship with a boss make people sick? *Our workshop participants answered with an overwhelming yes.* Many of them have had managers who made them sick, or they have seen it happen to colleagues. They tell us about a wide variety of stress symptoms ranging from vague psychological distress—can't sleep at night; feelings of dread, panic, or depression—to physical symptoms like stomach problems and headaches. And they are convinced that their managers are the cause.

How much of illness-inducing stress can be controlled by a manager? What do managers do—intentionally or otherwise—that contributes to employee illness? Although descriptions of the management behaviors that are most often responsible for employee illness cover the widest possible range, a few descriptions keep surfacing again and again.

People tell us a bad boss can "make" them sick by:

• being unpredictable
• whittling away at their self-esteem
• setting up win/lose situations
• providing either too much or too little stimulation on the job

---

Unpredictability

In any endeavor, the blow that comes when we least expect it is the one that hurts the most. Humans like predictability; it indicates that we have some measure of control over what happens to us, that we won't receive a random blow from the side that we think is safe. Predictability doesn't have to mean boredom or stagnation; it means that in most situations we know that a certain action will produce a certain outcome. Unpredictability, on the other hand, can cause great stress. Every time the boss walks in, a chain of thoughts and accompanying physical sensations is set off: Is he in a good mood? Will she yell at me? Should I ask him for help on this project, or will he bite my head off? The uncertainty triggered by an unpredictable boss can very often result in upset stomachs, headaches, nervousness, and other stress symptoms, although this may not be immediately apparent. Employees may continue to function for a long time with a budding ulcer or an impaired immune system, but productivity will suffer as they become more susceptible to minor illnesses.

In extreme cases, an unpredictable manager can completely disable an employee by creating an atmosphere of tension and uncertainty. Consider "Beth," who told us that the stress caused by her manager made it impossible for her to work:

Beth was excited about her new job with a major airline. Nervous about working in a formerly all-male environment, she was pleased to find out that her immediate supervisor was a woman. But she soon discovered that gender doesn't make any difference when a manager is unpredictable.

> I guess it started out as one of those woman-to-woman rivalries. In a male-dominated company, Gina (my manager) and I were both out to perform, to prove to everyone that we could do it.
> I didn't expect to be coddled, but I was totally

unprepared for her icy stare and her utter lack of warmth. When I gave presentations, she never encouraged me; she just crossed her arms and wore a blank gaze. When I asked how I was doing, she would say, "You're doing okay. If you're not, you'll be the first to know."

I soon found out, though, that Gina wasn't cold all the time. She had a red-hot temper, and she used it on me freely. The problem was, I was never sure when to expect an outburst: sometimes she let major mistakes go by without a word, and other times she'd blow up over things that I considered unimportant.

Before long, just the sight of her heading for my desk made my heart start to pound and my palms get sweaty. I'd get an urge to get up and head for the coffee machine, anything to keep out of her way!

Of course, I hated being chewed out, but sometimes it was actually a relief. At least when she pointed out my mistakes, Gina was helping me learn. I guess it wasn't the scoldings I minded so much as the fact that I never knew when they were coming. I became very nervous about making a mistake, so I double-checked everything and ended up behind schedule most of the time.

I tried to leave my work problems at work, but I found myself lying awake at night wondering just what I had to do to please her. I started getting migraine headaches, which kept getting worse until I couldn't concentrate at work. I had to take some time off to try to sort things out. Gina, as usual, was unsupportive. She implied that I was weak and childish, that I couldn't handle my job. I began to think that she was right.

Luckily, Gina soon was promoted out of the

49

department, and I came back to work. I was afraid my headaches might start again, but now that I had a new supervisor I seemed to be able to handle things better. I still make mistakes occasionally, but I don't feel incompetent any more. Now I finally feel that I'm working up to capacity.

Beth found out the hard way what happens when a manager sets an employee adrift in a sea of ambiguity. Gina was "freezing" Beth—withholding recognition for a job well done, keeping her at arm's length.

As a result, Beth never knew where she stood. She didn't know what was expected of her, and she didn't know if good performance would be recognized. All she knew was that she had a constant threat hanging over her head: "If you mess up, you'll be the first to know."

But what Beth objected to most was Gina's unpredictability—the "zaps" that came when she least expected them. The combination of freezing and zapping can be deadly. First the employee is demoralized by lack of feedback, then immobilized by punishment that comes like a bolt from the blue.

Beth reacted to Gina's freezing/zapping in two ways. First, she was so afraid of making a mistake that she became timid in her work. She double-checked everything, which slowed her down so that she missed her deadlines.

Second, the stress she felt from Gina's freezing/zapping caused migraines that forced her to take time off from work. Beth's health and productivity were so severely affected by Gina's behavior that she was virtually useless to the company until she got a new manager.

Whittling Away at Self-Esteem

In 1975 things were looking up for Deputy Sheriff Henry McKenna. He was finally back on the job after two years' recuperation from an automobile accident, and life was going well. Before long, however, McKenna noticed that morale in the department was low. Disturbed by what he felt were unusually severe disciplinary actions by a fellow officer named T. J., he took action by writing a memo to his superiors.

Shortly afterward T. J. was promoted to captain, with McKenna reporting to him. That's when McKenna's troubles began. First, he was transferred from a public relations assignment he enjoyed to patrol duty. Next he was bounced from day to night shift, then passed over for an expected promotion despite his seniority. Worst of all, T. J. frequently criticized McKenna in front of others for offenses McKenna believed were trivial or nonexistent.

McKenna's mood plummeted. Feeling himself the victim of a personal vendetta, he became acutely depressed. Eventually he caved in under the mounting pressure, quit the force, and filed a workers' compensation claim for psychiatric disability due to harassment by a superior.

The story above, taken from a workers' compensation case filed in a West Coast state, shows how

51

much damage a manager can do by whittling away at an employee's self-esteem. By giving McKenna poor assignments, passing him over for a promotion, and criticizing him in front of his peers, T. J. was constantly reinforcing the message: "You're incompetent. Your work is no good, and your feelings and desires don't matter."

In extreme cases like McKenna's, continually whittling at self-esteem can cause employees to quit or to perform so poorly that they are eventually fired. In other cases, employees may stay on the job; they're not fired, but they are "fired upon" by a boss taking potshots at their egos. In either case, both the employee and the manager lose. The employee loses self-confidence and a feeling of competence, and the manager loses whatever contribution to productivity a healthy, well-adjusted employee can make.

Low self-esteem isn't just an "ego problem"; in many studies, it has been linked to poor health, premature aging, and even death.

George E. Vaillant, M.D., recipient of a Research Scientist Development Award from the National Institute of Mental Health, writes of a four-decade study of 204 men that shows the link between health and self-esteem. The subjects' mental health was assessed periodically through adolescence and early adulthood into middle age. Those with poor emotional adjustment experienced more physical deterioration between the ages of forty-two and fifty-three than those who were better adjusted. Those who were chronically anxious, depressed, and emotionally maladjusted were more likely to show signs of early aging and an irreversible deterioration of health.[2]

George Engel, Professor of Psychology at the University of Rochester, studied 275 case histories of sudden deaths as well as many historical accounts of deaths of famous people and found four main categories of death-dealing sudden stress, one of which was "extreme sense of failure, defeat, disappointment, humiliation or loss of self-esteem."[3]

Another way that managers erode employees' self-esteem is by failing to "give credit where credit is due." Some managers take credit for work done by employees, while others give all the accolades to one person for what has been a team effort. It doesn't matter whether managers do this consciously or not; the net result is the same—lowered self-esteem and elevated stress.

Creating Win/Lose Situations

What has slipped away for many managers and executives is not just a sense of supremacy ("America as #2") but a sense of control. That is what they find so unsettling, so frightening, so frustrating, so intolerable. They feel at the mercy of change or the threat of change in a world marked by turbulence, uncertainty and instability, because their comfort, let alone their success, is dependent on many decisions of many players they can barely, if at all, influence.[4]

Control: it's a loaded word. For a manager, having control means being able to limit surprises. If you have your department "under control," you won't be caught unprepared. Often, control seems to be a clear-cut issue: Either you have it, or someone else does. Any time you grant (or are forced to give up) control to your workers, you may feel you have "lost."

This widespread concept of control sets up a "win/lose" mentality. If I (the manager) win, you (the employees) must lose, and vice versa. When both manager and employees start thinking this way, they become enemies. Work becomes a battle, where each side is intent on winning or at least forcing a stalemate—with predictably negative results for productivity.

When management "wins" a battle, employees naturally feel like prisoners of war. They feel defeated and subject to punishment. This feeling of being trapped can lead to a condition researchers call "learned helplessness." In experiments with dogs, for instance, pretreatment with inescapable electric shocks impairs the animals' ability to learn new tasks. Even when they are unleashed so they can avoid further shocks, they sit passively and endure the shocks rather than take action to escape.[5]

Unfortunately, many of us are taught the win/lose mentality from infancy. In almost any sports contest, the most important question is "Who won?" Even the very basis of our government—the idea of democracy and majority rule—rests on the idea that the majority "wins" over the minority by sheer force of numbers.

> ## Potential Outcomes of Win/Lose Situations
>
> - The atmosphere becomes distrusting, competitive, and hostile
> - Time and energy are diverted from the main issues
> - Creativity, sensitivity, and empathy are stifled
> - Authority conflicts become more frequent and bitter
> - Defensiveness, anger, and disenchantment are encouraged
> - Important organizational decisions are increasingly made by an isolated elitist group
> - New ideas are discouraged
> - The stage is set for sabotage by the "losers"
> - Deadlocks are created and decisions are delayed
> - Non-aggressive people are discouraged from participating

The "win/lose" concept may work for sports and government, but when it is used at a worksite it can quickly create resentment and alienation.

Providing Under- or Overstimulation

Pep talks, encouragement, nagging—whatever you call it, most bosses realize that their people need stimulation. It can spur them on to greater achievements—or push them over the edge into exhaustion. Quality, quantity, and timing of stimulation all help to determine whether it's a positive or a negative force.

Successfully providing stimulation involves more than just "lighting fires" under your people—a management behavior called "arson." Arsonists run from one crisis to the next, deliberately setting fires for people to fight. Under this sort of crisis management, employees always have too much to do and too little time to do it.

Providing the proper amount of stimulation is a question of energy and adaptation. Consider, for instance, two world-class runners: sprinter Carl Lewis and marathoner Frank Shorter. Each must run faster than the competition in order to win races, but their energy expenditures differ drastically. Sprinting requires a short-term all-out effort similar to the type of energy required when you tell one of your people to prepare a report for a meeting that will take place in two hours. Running a marathon, though, can be compared to working on a six-month project: The best way to finish it is to set a work pace that can be maintained for a long time, then move steadily toward the goal.

Athletes like Lewis and Shorter have the advantage

of years of experience and training. They've learned how to pace themselves in order to win. Your people, on the other hand, may not have that sort of experience and perspective. They rely on *you* to set a pace that enables them to be winners.

Learning how to set a proper pace for your people requires determining where their stress *comfort zone* lies. Research has shown that up to a certain point, performance increases with increasing stress; this is the point where it's effective to "push" people with deadlines, incentives, or other forms of pressure. Beyond that point, though, increasing stress causes performance to drop sharply, and only through *decreasing* the pressure will you be able to improve performance. This relationship between performance and stress was shown over seventy-five years ago by Robert M. Yerkes and John D. Dodson at the Harvard Physiologic Laboratory, and the results still hold true today.[6]

But *too little* stimulation can be just as great a problem as too much. In one study, participants were restricted to a single room and told to do absolutely nothing. Most of them rated the situation "unbearable" after three or four days. During the experiment, the bored subjects developed tension, sleeplessness, personality changes, reduced intellectual performance, and feelings of depersonalization. Happily, all the symptoms quickly disappeared when they resumed their normal activities.[7]

Other research shows that people overwhelmingly agree that boredom on the job is even more uncomfortable than long hours, heavy work loads, and pressing responsibilities. According to a study of two thousand people at the University of Michigan's Institute for Occupational Safety, those who reported being bored at work felt that their abilities were not being used and that their jobs did not provide as much complexity and variety as they wanted.[8]

Managing for health and productivity, then, means establishing enough stimulation to keep your people interested and challenged, but not so much that they will crack under the strain.

STRESS: A FORCE FOR WELLNESS OR ILLNESS

When a manager behaves in these ways—being unpredictable; eroding people's sense of confidence and self-worth; placing people in win/lose situations; or providing under- or overstimulation—employees are often subject to unnecessary stress that can make them ill. Let's look at the physiological side of stress to see how it affects health.

Stress means different things to different people. To a laborer, it is hoisting a heavy rock; to an engineer, it is the shearing force that is applied to an object. Yet all definitions of stress equate it with *tension*.

For most of us, stress refers to the psychological tension we feel in response to change. Although we may talk about "eliminating" stress from our lives, this is impossible; some sort of stress will always be present because change is ever-present. That's what work is all about.

Are You Managing for Wellness?

Stress itself is neither good nor bad. In the right quantities and the right situations, stress is a powerful force for growth and accomplishment. Stress is present in dissatisfaction with the status quo, the "itch" that spurs people on to new inventions or better ways of doing things. When stress can be managed, it can be used to accomplish great things. On the other hand, sometimes it can manage you; in which case it sets off a host of physical and mental reactions.

The best way to understand stress is to think of it as a force that elicits a response. Everyone's life is filled with potential "stressors," the sum total of life changes, events, and circumstances which occur each day. By themselves, they are not stressful; they only become stressors after we have interpreted them. Each of us has a mental filter through which we assess and label each event; we decide if any circumstance is going to cause stress for us.

When we label something a "stressful event," a whole chain reaction of physical changes is set in motion. We call this physical reaction the "fight-or-flight response"; it is what helped our ancestors protect themselves in a world where most stress came from physical danger.

Most of us can identify the physical reactions that form the fight-or-flight response. The last time you were in a near-miss auto collision, you probably felt at least some of these sensations:

• heart rate, blood pressure, and breathing rate increase

- pupils dilate
- blood rushes to muscles, skin becomes cold and clammy
- profuse sweating, heart palpitations
- sudden alertness
- neck and shoulders braced for an attack

These physical symptoms are the result of a complex chemical reaction that takes place almost instantaneously in the brain and bloodstream. Hormones including adrenaline are released into the blood and carried to every part of the body in less than eight seconds. At the same time, a message goes through the nervous system alerting the heart, lungs, and muscles to be ready for action. All this happens without any conscious thought.

THE POWER OF THE MIND

The fight-or-flight mechanism is so well developed that it doesn't require an actual physical event to set it off. Just visualizing a stressful situation is enough to start the chemical reactions we have just described.

The power of the mind is one of the biggest factors in dealing with stressors. It doesn't matter whether an event is actually threatening or dangerous; if we *label* it as threatening or dangerous, we will experience the physical reactions of stress and suffer the health consequences.

Fear isn't the only thought that activates the body's stress response; it may also be set off by negative thoughts. Visualizing failures or problems, dwelling on self-doubts or humiliating situations, feeling "trapped" in a job or a relationship—all trigger the same chemical responses as an actual physical threat. This chemical response causes the jittery, "edgy" sensation that we feel when we're in danger.

But the difference between responses that are set off by *physical* threats and those that come from *mental* stimulation is that the body tends to recover from the physical threat faster. A near miss in an automobile may set off a powerful fight-or-flight response, but as soon as you see that the collision has been avoided you begin to calm down. In twenty minutes or so, your body has "forgotten" that it was ever aroused.

Mental stress is another matter. Negative thoughts that plague us at work tend to linger in the back of the mind: Am I working up to par? Will I finish this project on time? Will I get this promotion, or will it go to someone younger? We don't "forget" about our problems at work; we tend to worry about them, thinking of them off and on throughout the day. This constant mental pressure sends a continuous message to the brain, which in turn keeps the body's stress response on alert almost constantly. What was intended to be an occasional chemical boost to escape physical danger becomes the body's constant state.

Physiologically, this means that the level of catecholamines (stress-related hormones) in the blood remains elevated far longer than it should, a condition that has been linked to internal organ damage. People whose catecholamine levels fail to decrease rapidly after high-stress situations show more "wear" on their organ systems. The body's failure to readjust chemical and hormone balances to lower resting levels after high-stress periods can lead to fatigue and more serious health problems.[9]

ENERGY DIRECTORS

Being a manager is a tough job. Causing stress for people goes with the territory; in fact, managers have sometimes been called "stress carriers," a descriptive but negative term. We'd like to replace "stress carrier" with the more positive "energy director." You receive pressure from above—you have deadlines, areas of responsibility, and criteria for performance. Your job is to translate that stress into positive energy, then direct it to the proper places in appropriate amounts. When used properly, stress is just another tool you use to get the job done.

In asking the question "Are you managing for wellness?" we start with the assumption that people want to do a good job. They generally don't come to work in the morning thinking, "Today I'm going to perform badly"—although they often end up doing just that. What happens is that their initial enthusiasm gets sidetracked by the way they are treated on the job.

We believe that frustrating people's desire to perform well—whether through misguided policies, lack of management skills, or an atmosphere that discourages wellness—causes unnecessary stress that's harmful to productivity and health.

PERKS BUILDS HEALTH FROM THE INSIDE OUT

So far, we've been looking at the negative side of the health/productivity connection. We've examined how "bad" management can cause illness-inducing stress. Now let's turn to the *positive*, and look at some ways in which individual managers can create a healthy and productive atmosphere.

In conducting our seminars, we've also asked participants to think of the *best* boss they've ever had. What were the management characteristics that made this person a pleasure to work for? Their answers contained many common elements—elements that correspond closely to the five parts of PERKS.

In fact, a number of studies support the link between employee health and the principles behind PERKS. On the negative side, lack of recognition for a job well done, boredom, and poor relations with co-workers are more often reported by people with coronary heart disease than by those with healthy hearts.[10] Job ambiguity—not knowing what's expected—has been linked to elevated blood pressure.[11] On the positive side, a high degree of job satisfaction is considered a major factor in promoting longevity.

Every manager at every level can use the five principles behind PERKS to help people "do better." That's the beauty of the PERKS system: No matter what your position in the organizational hierarchy, you can use your influence to help build health and productivity from the inside out.

The next five chapters describe the PERKS components in more detail. But for now we'd like to give you a "sneak preview" of Participation, Environment, Recognition, Knowledge, and Style.

PARTICIPATION: BUILDING THE FEELING OF BELONGING

Across the country, organizations are realizing the value of participation. Some companies are instituting formal programs to involve employees in the decision-making process; examples include Motorola's "Participative Management Program," Data General's "pride teams," and Honeywell's "quality circles" and "positive action teams." Besides improving morale and productivity, these programs serve as signs that American corporations are learning the value of employee involvement. But even if your company doesn't yet offer a formal program to encourage participation, you can involve people in many of the decisions that affect them.

First, why do people need to participate? Some managers remain unconvinced that employees want or need involvement in the decisions that affect them. Yet we have seen time after time that participation is vitally important to employees, on both an emotional and an intellectual level.

Emotionally, people have a deep-seated need to be "one of the gang." From our first experience of being chosen for a sandlot baseball team or being asked to the prom, to being included in the group when the office staff goes out to lunch, we want to belong. Being included in the decision-making process increases loyalty and a sense of belonging; it is an antidote for the win/lose mentality that can be so destructive on the job.

Intellectually, today's workers expect to be involved in the decisions that affect them. Younger employees grew up in families, schools, and neighborhoods that had shifted radically from their parents' experience. Television, working mothers, the "open classroom," all taught today's workers to expect a great deal of involvement in decision making. A survey of 2,300 *Psychology Today* readers found that respondents overwhelmingly asked for more control over the decisions that affect their jobs.[12]

Working Well addresses the *process* of participation: What can you do to make employee involvement a part of your work group? The chapter on *Participation* offers practical suggestions that can help managers at every level bring employees into decision making, and reap the advantages of full participation.

ENVIRONMENT: THE IMPORTANCE OF SHARED VALUES

You'll feel right at home in our small company, located in convenient, modern surroundings. Our small teams of professionals emphasize individual contributions and encourage creativity and initiative. We offer opportunities for growth and development along many paths. We reward people who do a tough job well.

By "work environment" we mean the *shared values* that management, employees, and stockholders have in common. For an ideal worksite those values might include support for risk taking, creativity, openness, fairness, a commitment to excellence, and a team spirit. Of the "shared values" that build a healthy and productive workplace, three are particularly important: trust and mutual respect, appreciation for people's unique abilities, and a commitment to safety and health.

The first key to a supportive work environment is that it is responsive to the needs of the individual; it provides opportunities for growth, and the security to know that risk taking will be rewarded, not punished. Such an environment is characterized by open communication, fairness, and consistency.

The second important characteristic is an appreciation of the unique qualities of individuals. An environment where individuality is respected gives people more opportunities to perform. People feel inspired to be creative in their approaches to problem solving and know they don't always have to do it "by the book."

Third, the ideal environment emphasizes safety and health. While this includes observance of safety rules, it goes far beyond that to encourage norms and group practices that actively support a healthy, wholesome atmosphere. This means being willing to confront some issues that can polarize the work force and cause productivity to plummet; for instance, the issues of smoking and substance abuse on the job.

Working Well views environmental improvements not as "extras"—nice touches that can be added when there is time—but as essential to the health and well-being of organizations and their people. In the chapter on *Environment* you will learn how to build a healthy supportive environment that not only makes people feel good but improves the organization's "bottom line."

RECOGNITION: IT STARTS WITH SELF-ESTEEM

There's always enough credit to go around.

One of the best things about recognition is that it's always available, and most of the time it's free. The pat on the back, the congratulations, the few words of praise in front of peers and superiors are rewards that can be used by every manager. Perhaps because recognition is so readily available managers tend to underestimate its power.

For an example of what recognition can do, think of a day when you left work feeling on top of the world. What had happened to you that day that made you feel so good about yourself? Chances are that sometime that day your boss had let you know that your work was noticed and appreciated. Recognition for one's efforts is essential for building self-confidence and self-esteem.

The power of recognition is evident among people who are lifelong high achievers. They have picked up their internal motivation from years of accumulated messages that have reinforced success. Since childhood they have been praised for their successes and in turn developed a "winning attitude."

Many underachieving employees, on the other hand, have been bombarded with information that leads them to believe that they are hopeless, hapless, helpless. Each failure—or unrecognized success—reinforces the message that they are "losers." The downward spiral is reversible, but changing these "losing" attitudes requires a steady diet of positive feedback.

It's easy to recognize proven winners. Employees who succeed at most of their efforts get most of the praise, most of the rewards. But what about the steady performers whose efforts are valuable but unspectacular? Recognition may be the key to elevating their performance. And for those employees for whom successes are few and far between, recognition can begin building the attitudes that lead to the behavior you want.

In the chapter on *Recognition*, you will learn how you as a manager can better recognize the achievement of your people, and how organizational efforts can help in that process.

KNOWLEDGE: THE ANTIDOTE FOR UNCERTAINTY

When my boss lets me know what's happening from the start, I really feel like it's *my* project.

Lack of knowledge can be one of the greatest stressors on the job, particularly in times of change. We all need some measure of order and predictability; we need to know that action X produces result Y.

When we don't have knowledge about important actions that affect us, we often create "worst-case" scenarios, allowing our thoughts to dwell on unrealistic fears. If we hear rumors of a reorganization in our department, we think, "There goes my job!" A sudden increase in closed-door meetings among top executives of a company sparks all sorts of speculation—most of it unfounded, and most of it pessimistic. These fears create unnecessary stress and cloud our thoughts, interfering with our ability to focus and concentrate.

Uncertainty—not knowing when the other shoe will drop—is one of the most stressful situations humans encounter. People do better when they have information about their jobs: what's expected of them, how to perform well, and how they will be evaluated. Knowledge is also an important tool in helping people grow within an organization, in staying healthy, and making positive lifestyle changes. It helps individuals move closer to their personal and professional goals and become effective contributors to the larger purpose of the organization.

Perhaps the most important knowledge the organization can impart is that we're all in this together; people working together can change negative norms, reach their goals, and create environments that support health.

In the chapter on *Knowledge* we'll discuss what kinds

of knowledge people need, and how managers and organizations can make the necessary knowledge available.

STYLE: PROVIDING SUPPORT AND DIRECTION

If employees get the same response time after time to all their actions, how do they know when they're getting better? How do they know when they've made a mistake?

That's where management style comes in. Choosing the right management style makes the most of people's skills and interests, inspires them to perform well, and sustains their good performance. Most importantly, management style isn't something you do *to* people, but an interactive process of analysis, communication, and flexibility.

The right management style is necessary not only for top performance, but also for maximum employee wellness. A style that provides too much direction will have employees "champing at the bit," eager to perform but held back by a manager who doesn't allow them full rein for performance. The result will be frustration and resentment—and all the health problems that can cause.

Too little support, though, can be just as bad for health. The "leave-alone" management style is the most widely used among managers: They tell people what to do, then leave them alone to do it. For inexperienced employees or those trying a new task, being left alone is frightening and stressful. Not knowing what to do either paralyzes them or causes them to make mistakes. Then when the manager discovers the missed deadline or the botched project, he descends with a sharp reprimand or punishment—the "zap." The result is that employees become completely demoralized, timid, and afraid to try anything new.

A manager who consistently uses the wrong style can cripple an entire department. Productivity drops because people don't get the support or direction they need to perform well. They constantly hover on the edge of illness, victims of headaches, digestive problems, and all the minor but chronic aches that go with stress.

In the chapter on *Style* you will learn that choosing the right management style is what ties the rest of the PERKS together. Delivering the proper amounts of support and direction when needed can help people succeed at the tasks you set for them.

SOME ADDITIONAL THOUGHTS ABOUT PERKS

In the chapters ahead, we'll show how an individual manger can become a positive force for health. You will find practical "how-to" advice for using Participation, Environment, Recognition, Knowledge, and Style in day-to-day management to help your people be the best they can be.

Whenever possible, we've applied the PERKS principles to organizational efforts designed to improve employee health. For managers who are in a position to establish or support health-promotion efforts, you'll find success stories and helpful guidelines well marked throughout the book.

Chapter 3
Participation

> People do better when they are involved in the decisions that affect them.

Organizations have the capacity to solve most if not all of their own problems. In other words, organizations have the potential to be self-healing. The problem is, companies often don't use the resources that they have. They get locked into roles and stereotypes that limit their flexibility: "Managers solve problems"; "People closest to the problem don't have the objectivity to solve it"; "We have to go through the proper channels"; "We'll need to bring in an outside consultant."

These stereotypes become self-fulfilling prophecies. Constant use of consultants gives employees the message that their input isn't valued, that they're somehow not "good enough" to contribute to solutions. When managers don't seek or welcome their help, employees eventually learn not to offer it. Then managers say, "They don't give us any help, therefore they must not know anything," and the cycle goes round and round.

Tom Peters, co-author of *In Search of Excellence*, says, "Too many managers treat employees like children—and then are surprised when they behave like children." Successful business leaders, Peters feels, have one thing in common at the end of the day: They share a bone-deep belief in the intelligence, creativity, and ability of the people they employ.

Participation starts with positive assumptions about human nature and the capacities of the people that you hire. The next step is to turn those assumptions into action by providing opportunities for people to get involved in their own destiny as well as the destiny and success of the organization.

Why don't all organizations involve their people in decision making and problem solving? Why do many managers cheat their people out of opportunities for becoming involved in meaningful work and decisions which affect them? Because they're locked into old assumptions and old management systems.

RESPONSIBLE VS. RESPONSIVE: A FEW LETTERS MAKE A BIG DIFFERENCE

Most organizations are structured as pyramids. The chairman of the board, the president, the board of directors, and the other "important" people are at the top, the middle managers and supervisors in the middle. Who does that leave at the bottom of the pyramid? The hourly employees, the people who deal most often with the customers and who do the physical work of the organization.

There's nothing wrong with a pyramidal structure; most organizations are built that way. The problem comes when people start to think like pyramids. When you think like a pyramid, each level of the hierarchy works for the level above it.

That's the way most people in organizations think, and it reinforces the belief that managers are *responsible* for the actions of their employees and the success of their departments. Managers are the ones who make the decisions. Employees, in this way of thinking, aren't responsible—their job is to be *responsive* to managers.

So people quickly learn that pleasing the boss rather than doing quality work is their primary goal. In many companies we've seen people working late only because they've got a boss who never seems to go home. Whether the demands of their jobs require long hours or not, they know that staying late will convince the boss that they're working hard.

75

Under the traditional pyramid structure, employees lack responsibility for solving their own problems. They become helpless, shifting all their problems onto the manager's back. And that's no fun if you're the manager! Your work load gets larger and larger as you take on the problems of the whole work group. Everyone suffers.

Think for a moment, though, what would happen if we philosophically turned the pyramid upside down. Now the president and the top managers are at the bottom; at the top of the organization are the hourly employees, with the customers somewhere above them. Employees now have the responsibility to work well and solve problems; managers are responsive to the needs of their people, giving them support in their problem solving. The manager no longer carries the whole work load but provides whatever aid and support employees need.

When we turn this pyramid upside down, it changes the whole structure of who works for whom: The traditional bosses now work for their people. This means that the only purpose of each layer of management is to respond to the needs of all the layers above them. Suddenly, communication between layers and involvement of all employees in decision making take on a new significance.

DON'T "CHEAT" PEOPLE OUT OF CHALLENGES

What is the most important part of a manager's job? To manage people and problems? No, the most important function a manager can perform is to *give people opportunities to manage what they can by themselves.* A manager who takes on responsibility for all the problems in the group cheats employees out of countless challenges and opportunities for growth, denying them the chance to have more interesting jobs.

With "reverse-pyramid" thinking, though, managers no longer have to assume responsibility for all the problems that may arise. Instead, their goal is to help people closest to the problem find ways to solve it themselves. How many managers, for instance, would ask their secretaries to redesign office procedures? Yet secretaries are the ones who deal with those procedures daily. When a problem arises, the first step in handling it should be to ask the people most affected by it, "What are your ideas? How can we make things work better?"

A new generation of employees at all levels have begun to expect challenges from their jobs. As the average education level among workers has risen over the past thirty years, people have grown to expect that their opinions and contributions on the job will make a difference. When those expectations are not met, the result is frustration, boredom and unnecessary stress.

TAKE CARE IN THE BEGINNING, FOR THE END WILL TAKE CARE OF ITSELF

Are you looking for a "quick fix" for your organization's problems? Do you "want it done yesterday"? Unlearning old ways of management takes time, and setting up the structures for participation seems too slow for many. Yet a participative workplace is the basis for the PERKS system. Participation lays the foundation for the other four components—a healthy environment, recognition, knowledge, and management style.

In order to understand what we mean by "participation," let's take a look at two extremes of a management spectrum. These are hypothetical situations; it's unlikely that any real-life work situation would contain all the elements of either model. A totally directive, autocratic system is illustrated by the comments in the left column, and a democratic, participative model is shown on the right.

AUTOCRATIC	DEMOCRATIC
Management has little confidence or trust in subordinates.	Management has complete confidence in subordinates.
Employees are seldom involved in any aspect of decision making.	Decision making occurs at all levels but is well integrated by management.
Decision making and goal setting are top-down activities. Orders are issued down the chain of command	Communication flows freely—up and down the hierarchy and among peers.
Superior/subordinate interactions are characterized by fear.	Interaction is extensive and friendly; people trust each other.
Employees work under the threat of punishment with only occasional rewards. They seek satisfaction of "lower-order" needs—good wages, job security, safety.	Workers are motivated by participation and involvement in developing economic rewards, setting goals, improving methods, and appraising progress.

Participation

AUTOCRATIC	DEMOCRATIC
Although the central process is highly controlled by top management, an informal organization generally develops in opposition to the goals of the formal organization.	Lower units are fully involved in the control process. The formal and informal organizations are one and the same. All social forces in the organization support efforts to achieve the stated organizational goals.[1]

Where does your company fall on the spectrum? Which method of decision making is the norm in your company, autocratic or democratic? Both types have their place, but they send vastly different messages to employees.

Autocratic change starts from the top of the traditional pyramid. The decision is made at an organizational level and management announces, "As of September 1, the following will occur." Organizations that use an autocratic style hope that as people get used to the change they will eventually develop positive attitudes toward the new behavior.

The advantage of the autocratic method is that it's quick; you can start it tomorrow. But the disadvantage is that people may resist it. They may undermine the change and try to "get even." More commonly, they will wait out the change, looking forward to the time when "we'll all be back to normal around here."

A democratic process, on the other hand, is a series of steps which slowly build commitment and enthusiasm. The advantage of democratic change is that people are committed to it. Their attitudes are positive, and they're behind the change all the way. The disadvantage is that educating people and changing attitudes can take a long time.

While it's sometimes necessary to use autocratic methods to get a change going in a short period of time, in reality it can be a false economy. Change can only be sustained over the long term when people have "bought into the idea" and made it their own.

In this section we'll apply the concepts of participation to one type of innovation: the effort to establish a health-promotion program. But whether the effort is health promotion or cost reduction, the introduction of new technology or any other change, the theories and methods of participation remain the same.

This chapter shows how to increase employee involvement. Participation, far from being a revolution, is a set of common-sense principles that can be followed in any workplace. To show how this works on a practical day-to-day basis, we have chosen five management techniques that we have witnessed: enlisting the support of "natural leaders," gathering facts before you start, responding to employee input, using incentives tied to performance, and developing your own "approachability."

IDENTIFY AND ENLIST NATURAL LEADERS

In any group there are key people to whom others turn for advice and guidance. There is undoubtedly at least one such "natural leader" in your work group whose support is crucial when you introduce a change. As in a political campaign, you've got to get this leader's endorsement if your people are to "buy into" the new idea. Employees who are involved from the beginning feel a sense of ownership in the decision and are more committed to seeing it work.

Often, though, what should be a democratic change is introduced to employees in an autocratic manner. The idea may sound so good, harmless, or natural that you think you can skip the time-consuming process of wooing the group's leaders and seeking support from the group as a whole.

This is often the case for health-promotion efforts. A well-meaning CEO, human resources director, or medical director will decide that offering health-promotion programs is a good idea. They get the required approval, then present the program to employees, expecting an enthusiastic response. When the new program is met with indifference or even hostility, the promoters can't understand why. They have forgotten the importance of involving employees *before* they implement the change.

When Mark was working as a medical director at a fast-growing electronics firm, he learned the difference between programs that come from the "top down" and employee involvement that comes from the "bottom up."

> The company was growing rapidly—so rapidly that we had run out of room in the main building and began renting space in trailers and warehouses. The physical separation and rapid growth led to a decentralized, "spread-out" feeling among many of the workers.

81

One group that felt particularly isolated was a crew of seventy assembly line workers situated in the basement of a supermarket several miles away. We decorated the temporary quarters with plants and bright colors, trying to make the workers feel that the company cared and that they were not forgotten. But no matter what we did, they still had a sense of being apart from the company.

We were offering various health-promotion programs and classes for all employees, but this was the one group that just never got involved. They seemed uninterested in any company-sponsored activities. Our human resources manager kept trying to come up with ways to increase their sense of connection, but they remained isolated.

One day, an overweight assembler from this group came to me, saying: "Please, Doctor, put me on a diet." Usually when you hear those words, you know that you are dealing with someone who has tried and failed many times.

"No, I won't put you on a diet," I told her, "but if you can get twenty people from your work group together, we'll develop a program that will help you get the support you need to lose weight."

Judging from her group's lack of response to previous health-promotion efforts, I expected never to see her again.

A week later she handed me the sign-up sheet of 32 names—almost half of all the employees in her group. We designed a program and the group worked together for the next ten weeks, exploring issues, walking together, learning about nutrition, and examining how they could support each other's new healthier habits. And I learned about health promotion through bottom-up employee involvement.

Why did the weight-loss program work when all the other programs had failed? For two reasons: first, because it was what the employees wanted and not what someone else thought was good for them, and second, because the leadership for the program evolved naturally out of the social structure of their work group.

Identifying and enlisting the support of natural leaders within a work group is one of the best ways to foster employee participation. Natural leaders are not necessarily supervisors or managers; they are people toward whom other employees naturally gravitate, people whose example they are willing to follow. To identify the natural leaders among your people, ask yourself the following questions:

To whom do other employees look for advice and guidance?

In meetings, which employees seem to reflect or sum up the feelings of the group?

What trait or value seems to be most prevalent, and who represents this best?

Ideally, you want the natural leader to support your policies, but even if you "agree to disagree," you have eliminated one potential source of bad feelings; you have consulted the natural leader and listened to the group's feelings before making a decision. The group may not agree with the decision, but they must acknowledge that you have involved them in the decision-making process.

Gaining support from natural leaders may be as simple as asking them. When planning a new program or other change, gathering opinions involves a few basic questions: "What do you think about this? How can we make it better? How can we get people to go along with it? How will they react to this?" By soliciting the support of natural leaders, you can make the grapevine work for you.

BEFORE YOU START, GET THE FACTS

Deciding on a participative style and enlisting the support of natural leaders are only the first steps to enhancing participation among your people. As in the health-promotion example, managers may sometimes misinterpret a lack of response as apathy when it really results from a failure to find out what employees really want.

Employees are most likely to "buy into" those programs and projects that fit their needs. The simplest way to find out what they need or want is by asking them, either informally or with questionnaires. The following sample shows the types of questions you would want to ask.[2] A good questionnaire may provide data that can help you identify a target group where there is a likelihood of short-term success. You

Participation

might also be able to find out how significant your opposition is—how many people are resistant to the new idea and to what extent.

Program Interest Questionnaire

Program Interests: The following are examples of the types of programs that might be offered. Please indicate by circling the appropriate response how interested you would be in participating.

1 = Very Interested
2 = Somewhat Interested
3 = Somewhat Uninterested
4 = Not Interested

An introductory program that explains how you can become all that you can be and avoid disease, and which would help you to design and plan your own wellness program	1	2	3	4
A stop-smoking program	1	2	3	4
A program on nutrition	1	2	3	4
A program on stress management	1	2	3	4
A program on exercise for good health to help you begin a personalized exercise program	1	2	3	4
A program on weight reduction	1	2	3	4
A dancing-for-fitness program	1	2	3	4
A program on accident prevention and safety	1	2	3	4

Program Interest Questionnaire (*continued*)				
A communications program to help you improve your human relationships	1	2	3	4
A recreation program (e.g., softball, basketball, volleyball)	1	2	3	4
If there was an opportunity to do so, would you want your family to participate in these programs?	1	2	3	4

IF YOU ASK, BE PREPARED TO RESPOND

It's not enough just to get the facts before you start; you have to be prepared to use them. Many organizations take surveys and hold open meetings to get employee input before a planned change. They behave as though they are going to take into account employee needs and wishes—and then disregard them in the final project. By building false hopes and dashing them, organizations breed sarcasm and cynicism in their people: "Oh, boy, another employee attitude survey. They might as well paper the restrooms with these; they don't listen to what we say anyway."

Ignoring employees' suggestions when you've asked for them multiplies the potential for anger and resentment. Consider the engineer in the story below, whose resentment is still seething even a decade after the incident he describes:

Participation

About ten years ago we had to redo our sample compositing room. It was all going to be modernized. They asked the eight or nine chemists who worked there to come up with a plan. "You're down there all the time," they said, "give us your thoughts." Everyone said, "This is our chance to design a workplace that really makes sense."

We had many meetings in people's homes. Everyone agreed that a sloped floor to the drain should be built so that you could hose down all the raw materials quickly and efficiently. Some of our people were very good at drafting, so we made blueprints of exactly what we wanted and sent it all to management.

You can guess what happened. They spent $45,000 to remodel that room and they put in a straight floor.

My enthusiasm for my company has been beat out of me. So now my enthusiasm is for me. I do a professional job. I can cover my bases. But as far as the fate of my company, as long as there is a signature on my paycheck every week, I couldn't care less—so don't tell me how they are interested in our suggestions for improvements.[3]

How can you avoid situations like this? In this case, the company may have had a reason for disregarding the group's suggestion—but they didn't explain it to the chemists. As a result, the design group felt that their input was ignored.

You can't act on every suggestion you receive from employees, but you can let them know that you have received and considered the suggestions. Acknowledge all contributions, even the ones you don't use. If you ask for input, as in the example above, share the reasoning behind the ultimate decision.

For instance, when management asks for suggestions for a health-promotion program, employees often suggest installing showers in existing buildings. Often management either ignores the suggestion or rejects it curtly: "That's too expensive." Employees may think that the expense argument is simply an excuse to turn down their suggestion. If, however, management presents the cost breakdown of building the showers versus the number of users, people will see the process that led to rejecting the suggestion. The net result is the same—the showers won't be built—but now employees don't feel shut out of the decision.

USE INCENTIVES TIED TO PERFORMANCE

In searching for means to encourage employee participation, growing numbers of companies are using financial incentives. Monetary rewards in the form of piecework or bonuses have long been used to encourage productivity, but now the idea is beginning to spread to other forms of participation as well. For instance, many organizations are using financial incentives to encourage participation in health-promotion programs.

The principle behind monetary rewards is simple: Employees are either rewarded or punished economically for certain health-related behaviors. Incentives may take the form of money—either in a lump sum or as periodic payments—or prizes such as gifts or company-paid trips.

Participation

One such incentive program is Hospital Corporation of America's "Aerobic Challenge." Designed to promote regular aerobic exercise, the program pays employees twenty-four cents for each "unit" of exercise they report. Different sports rack up units according to degree of aerobic difficulty:

UNIT EQUIVALENTS IN:
Running	1	mile
Swimming	¼	mile
Bicycling	4	miles
Aerobic dancing and Jazzercise	¼	hour
Racquetball	½	hour

Participants are paid for exercising only if they accumulate over thirty units per month. That way, the program encourages regular exercise, not "weekend athletes." In 1979, the first year the program was in existence, HCA paid out $2,833.10 to the approximately 30 percent of employees who participated in Aerobic Challenge. Other companies have experimented with financial incentives to achieve weight loss, not smoking, and other health goals.[4]

Financial incentives are just one way of rewarding desired behavior changes. For those who have the skills and motivation to change their behavior, a monetary reward can provide the incentive to keep them on the track.

Financial rewards, though, are usually individual rewards. For many people, money is not enough to make up for the lack of skills or self-confidence that stands between them and their goals. For the majority of employees, participation—and ultimately success—is easier when they have the support of a group all working toward the same goal.

This process of social support uses the powerful effect of peer pressure for the benefit of the whole group. The group provides encouragement and the knowledge that each individual's contribution is valued, and in return sets up a network of mutual obligation. Achieving the group goal depends on the contributions of each member.[5]

General Motors applied this knowledge about the importance of group goals in a program to change a serious health risk: the non-use of seat belts.

> Employees at the General Motors Technical Laboratories were encouraged to sign pledge cards promising to use their automobile safety belts for one year. Usage was monitored at the entrance to the company parking lot.
>
> The payoff was simple: if the group met a pre-specified usage goal, the company would hold a sweepstakes drawing from among the pledge signers.
>
> In the first six weeks following its introduction in 1982, seat belt usage increased from 45% to 70%. The company awarded a weekend holiday in Toronto, the use of a company car for a week, and 15 watches to the sweepstakes winners.
>
> Buoyed by their initial success, management raised the goals and the stakes. A second sweepstakes aimed at achieving 75% usage drew 85% of the employees into the program. The results: an 82% usage rate; the prize: a trip to Hawaii.[6]

Another example of a program using both financial incentives and group support, this time involving a weight-loss competition among three banks, was conducted by Kelly D. Brownell, Ph.D., of the Department of Psychiatry, University of Pennsylvania. In this case employees formed teams to enter the weight-loss competition, with each participant contributing five dollars to a pool. The team that achieved the greatest percentage of its total pound-loss goal (sum of the difference between ideal weight and actual weight for team members) won the entire pool of money.

The competition actually offered very little in the way of an organized behavior-change program. Instead, it relied on peer pressure that made success or failure of the group a matter of public record: weekly weigh-ins, media coverage, and lobby scorecards similar to the United Way thermometer.

After fifteen weeks, the participants had lost an average of twelve pounds each. What's more, at a six-month follow-up the average participant had maintained 80 percent of the original weight loss.[7]

WHAT YOU CAN DO: USING ENCOURAGERS

Individual and group incentives are two ways to encourage employee participation in any change effort. But a more basic means of building employee involvement is by monitoring the way you interact with them. Your choice of words and your body language can let people know that their input is welcome and valued — or it can scare them off. Consider the case of "Tom," the sales manager for a Southwestern high-tech firm:

Tom was a master at squelching enthusiasm. Whenever one of his people came to him with an idea, Tom would listen intently for perhaps thirty seconds, leaning across his desk and fixing the salesperson with a piercing gaze. Then he would interrupt with an impatient wave of his hand and deliver one of these "killer phrases":

"The problem with that idea is..."

"No way it will work here."

"Be serious, this is the real world."

"We've tried that before!"

Then Tom would rattle off four or five reasons why the idea wouldn't work. The salesperson would apologize for taking up his time and slink out of the office feeling like a fool.

Why did Tom pounce on his people? He was a very intelligent, competitive man with a high need to have control in his world. He regarded every encounter with his people as a battle of wits—a battle he usually won because of his intelligence, his experience, and his rank. What he didn't realize was that by winning the battles he was destroying his people's enthusiasm and self-esteem.

When his demoralized sales force started performing poorly, the company called in a consultant to try to solve the problem. The consultant immediately pinpointed Tom's combative management style as a major obstacle for the sales force.

With the help of some group meetings and role playing, the consultant was able to help Tom see how his style was eroding his people's enthusiasm and creativity. He helped Tom develop a new style encompassing body language, tone of voice, the use of appropriate questions, and "encourager phrases."

First, Tom would give the salesperson up to five minutes of undivided attention—without interjecting his own ideas. With the use of role playing and rehearsal, Tom learned to modulate his tone of voice, adopt a more relaxed posture, establish eye contact, and in general seem less threatening.

The consultant helped Tom develop a vocabulary of "encourager phrases" to replace his old "killer phrases." Now instead of jumping in with a negative comment, Tom would respond with phrases like these:

"Keep talking, it sounds good."

"That would be interesting to try."

"How can I help?"

"Let's make it work!"

The change didn't come easily. Tom posted notes all over his office to remind himself to use his new "approachable" style. At first he felt like a phony.

But he and the sales force soon discovered that the new style was working. His people knew that he would at least listen to them. Once he started listening, Tom found that they had some very good suggestions. The "approachable" style soon began to feel more natural, and Tom found that his sales force reacted with enthusiasm and creativity.

DEVELOPING YOUR OWN "APPROACHABILITY"

Using "encourager phrases" didn't turn Tom into a pussycat, but it did help his people get over their fear that he would immediately shoot down or take over their ideas. It's a common problem; it's usually easier to find reasons why a new idea will fail than to search out ways to make it work. When employees learn that all their suggestions will be rejected out of hand, they react by withdrawing. They feel alienated, left out of the planning and decision-making process.

One way to encourage participation is to work at becoming as "approachable" as possible. Like Tom, most of us slip into bad listening habits at times. When we're busy or preoccupied, we might be more brusque than we intend to be. But like Tom, we can consciously change our choice of words, tone of voice, and body language to make ourselves more approachable. Besides using "encourager phrases," try:

- Active listening. Show you are listening by punctuating the conversation with "uhmmm," "uh-huh," "go on" (but don't interrupt). Focus on important statements by repeating key words: "Improving quality?" "Increase 45 percent?" If necessary, restate what the other person has said; this will help both of you make sure you have understood.

- Body language—*yours*. Leaning slightly toward the other person with direct eye contact is the typical listening posture in our culture. A small hand movement with an open palm often encourages them to continue.

- Body language—*theirs*. Direct eye contact indicates interest or intensity; looking away can indicate boredom or discomfort. Watch to see

94

whether they shift toward or away from you as
you respond to their suggestions.

- Good use of questions. Use open questions
("What do you think?" "How can we solve this?")
to encourage people to talk. Use closed questions ("Will you be at the meeting?") if you want
only a short answer. Open questions make you
much more approachable; use closed questions
only if the speaker wanders off the subject. Avoid
questions that suggest an answer ("Wouldn't it
be better if...") or that start with "Why"; these
can put the answerer on the defensive.

GIVE UP A LITTLE CONTROL TO GAIN A LOT MORE

Adopting a more "approachable" style may not be
enough, though, without changing the basic underlying assumptions. Tom, for instance, rejected his subordinates' suggestions because, frankly, he thought he
was smarter than they were. True, he knew more about
the business than they did. He had more experience
and access to more sources of information. But that
didn't mean he had cornered the market on creativity
and intelligence. Admitting that he didn't have all the
answers would open him up to fresh viewpoints and
new ideas.

Instead, he had set himself up as the sole source of
knowledge and arbitration, the one with all the answers, the one to turn to whenever his people had a
problem. In the process he robbed his people of the
opportunity to make their own decisions and learn
from their own mistakes.

Entrepreneurs are particularly prone to this error. Trying to run every aspect of their businesses by themselves often costs them more in the long run than helping their people grow into positions of responsibility. In a 1984 *Harvard Business Review* article, Glenn H. Matthews describes how his decision to "build an organization" rather than "run a business" helped him escape from the entrepreneurial trap of trying to run everything himself:

> My job is getting easier and my subordinates are maturing as decision makers able to identify and solve problems as well as manage their own subordinates effectively. We are becoming an organization....
>
> Both running a business and building an organization can be compared to wrestling with a giant octopus. Your subordinates can hold its arms, while you sweep away the inklike cloud to see what's going on. It's better than being asked what to do about the octopus while you are drowning in its clutches. That can happen when you "run a business."[8]

Successfully "wrestling the octopus" in your own department means establishing an atmosphere where your people are encouraged to grow, even if it means failing sometimes. You can do this by:

- Admitting that you don't have all the answers. The importance of asking for help was underscored in an experiment at St. John's University where all professors in addition to their regular area of expertise were required to teach a class on one subject they knew nothing about. In student evaluations at the end of the term, almost all the professors were rated higher in the subjects they knew nothing about than in the subjects they normally taught. When they didn't consider themselves the "experts," they relied more on the resources of the class—with the result that the students were more satisfied.

- Letting your people know that they can take risks without risking their jobs. Usually the fear of what will happen if they fail is far worse than the consequences of failure.

- Becoming "approachable." Pay attention to your body language, tone of voice, and choice of words. While substance is more important than style, it's your style that either encourages or discourages seeing the substance.

THE LIFE CYCLE OF PARTICIPATION

Although everyone shares the basic desire to have a say in decisions that affect them, participation doesn't always come easily. You can't switch overnight from an autocratic environment to a democratic one; neither the manager nor the subordinates will have learned the skills and habits needed for a truly participative work climate.

In its beginning stages, participation takes much more time than autocratic management. Meetings are more frequent and more unwieldy as each member struggles to be heard on every decision. Employees unused to asserting themselves and making their own decisions will need more direction and support from you to help them grow into full members of the group.

At this point, a certain amount of regression is to be expected from everyone. It's not easy for a manager to change leadership habits that may have been years in the making; you know it's easier in the short run to make decisions yourself than to go through the struggle of reaching a group consensus. Your people, too, may backslide into old attitudes of cynicism or apathy, waiting to be told what to do. It takes time for the habit of participation—and the belief that what they say and think really does matter—to take hold.

At the beginning of this chapter, we stated our basic belief about participation: "People do better when they're involved in the decisions that affect them." Most workplaces provide far too little participation, leaving employees feeling alienated and apathetic. In most situations, employees and company alike can profit from allowing people a greater say in the decision-making process.

Although participation is in short supply in most work situations, we realize that there are times when it's not appropriate. Participation at the workplace shouldn't be a strict democracy; although everyone's input is important, the truth is that some people's viewpoints really do carry more weight than others'. Younger or less experienced employees might bring fresh ideas and creativity to their jobs, but that must be weighed against the proven records of the more experienced people. Those who have a sense of the value and history of the company and the scope of the industry, those who care about the company's long-term welfare—perhaps these people are slightly "more equal" than the others.

In light of these facts, we'd like to add one word to the ending of our statement on participation: "People do better when they're involved in the decisions that affect them...usually."

As a manager, then, you're more than a vote counter. You're part investigator, part cheerleader, part architect, and part salesperson. You gather input from your people, weigh it, and render a decision—always remembering that unless you have their buy-in, you won't get their wholehearted support.

Chapter 4

Environment

> People do better when the
> environment gives them
> opportunities and choices to
> perform well.

The organizational environment goes by many different names, most often the "work climate" or the "corporate culture." We like to think of it as the sea of organizational values, policies, and norms that surrounds people, influencing their sense of need satisfaction and opportunities to perform. Like the natural environment, it may be taken for granted much of the time, yet it still has a strong influence on how people live and work.

An organization's environment is shaped by a number of factors both external and internal. Company policymakers have little influence over many of the external forces that influence their environment: markets and competition, the condition of the economy, prevailing social norms.

Most organizations, though, make a conscious effort to use internal influences to shape their corporate culture; for instance, companies may attempt to shape their environment by molding their image through advertising and public relations.

Often in the attempt to engineer the kind of image they want, management may lose sight of the most important aspect of the corporate environment: their employees' perception of it. Regardless of the CEO's speeches to the press or the company's latest advertising campaign, what matters to employees is "how things really work around here." For employees, that spells out the impact of the environment on their health and productivity.

Recognizing this, many companies are using "work climate" or "organizational practice" surveys to find out how things really operate within the organization. Do employees feel the waters are friendly? Do they agree with—or even know about—the organization's mission and purpose? Do they trust the company and its representatives? Do they feel secure taking risks? The answers to these questions provide a more accurate picture of the corporate culture than can be obtained from a glossy brochure or a talk with the CEO.

THE ROLE OF TOP MANAGEMENT

Companies where the environment is safe, friendly, and nurturing often became that way through top-down leadership. Many companies take on the personality of their founder or their CEO, whose values permeate every level of the organization. The role of the CEO and other top managers is to constantly interpret and demonstrate the values of the organization. When the eighty-nine-year-old Chairman of the Board of Matsushita, one of the largest electronics firms in the world, was asked what he thought was the most important aspect of his job he answered, "To model love." As he saw it, the values of the organization passed through him. As a result, his job was to wander around the company and affirm people, to create an environment that showed people that they were important.

Telling people "You are important" may have little impact if it is only an isolated incident and if everything else the organization does sends out contrary messages. But when the message is repeated, and repeatedly demonstrated, it sinks in and takes root until employees internalize the feeling of importance.

Most of us can identify various points in our lives where these messages first took hold. For Margie, it was when she was an undergraduate at Cornell University:

Environment

> One of the things that our professors would tell us all the time is that our generation would one day be leading the country. That gave us a sense of responsibility; when we heard that over and over and over again, eventually we believed that it was true. We believed we were going to use the educational opportunities that we had. Now I see that it did make a difference: a lot of my classmates are now leaders in their fields.

There are many things you as a manager can do to build that sense of responsibility, that push toward success, in your people. While your job description probably does not include "modeling love," there are many steps you can take to build a climate of trust and mutual respect:

- Treat people as if everyone's time has the same value. By starting meetings on time you are demonstrating respect for the people who made the effort to arrive promptly, instead of rewarding the latecomers by waiting for them.

- Use a listening exercise suggested by one of our colleagues for showing mutual respect at meetings. He hypothesized that the reason his meetings were taking so long was that people weren't listening to one another. They were busy formulating their own responses while someone else was talking, then after the meeting they stood around in the hallway and asked each other what had happened. So he redesigned the meeting procedure. When a speaker was finished presenting an argument, they would move clockwise around the table until they found someone who wanted to speak next. That person would have to start by saying the name of the previous speaker just to show that he had listened and was now ready to respond.

- When you must discipline people, do so tactfully. Choose a discreet time and place, rather than in front of others or in a meeting. Feeling humiliated is a major cause of stress and illness; try to correct people without embarrassing them unduly. The proper way to deliver feedback—both negative and positive—is discussed more fully in the "Recognition" chapter.

- Promote from within rather than always hiring from outside. One of the best ways to demonstrate to employees that they are important is by providing a career path through the organization. When the company adopts a policy of promoting from within, employees know that their own professional growth is encouraged and expected. One way to keep people aiming at growth is through "skip-2" meetings, where employees meet with someone at a level above their own manager. That way employees gain a broader view of the organization, its goals, and where their own growth might lead.

- Emphasize solving problems, not placing blame. Things go wrong even in the best-managed companies; but the real mistake occurs when managers and employees put a great deal of effort into finding out "who messed up" rather than "how can we correct this?"

APPRECIATE UNIQUENESS

People do better when they know that their contributions are important. Yet most of us tend to like and reward people who are like ourselves. Even the best-intentioned manager is likely to be more sympathetic to those whose ideas are in tune with the manager's. But those who are different can often make valuable contributions. A consultant we know saw this in a sales group:

They called themselves the Jonathan Livingston Seagull Seven. They liked to think that they were not just an ordinary marketing team—they were creative people with soaring spirits.

Actually, they should have called themselves the Jonathan Livingston Seagull Six—plus one. Ralph knew he didn't fit into the group, and everyone else knew it too. They thought he was stodgy, and he thought they were flaky. He felt it his duty to point out to the other six when they were letting their fantasies carry them a little too far.

When Ralph first joined the group, he offered his suggestions freely—and the rest of the group pounced on him. "Don't be such a pessimist, Ralph," the others said. "Let your imagination go. Otherwise, you'll never get anywhere."

But Ralph just couldn't stop being practical. He soon found that the other members of the group ignored his comments at meetings and continued their fantasizing. Soon, Ralph began to doubt himself. Maybe they were right; maybe

he was too earthbound. He became timid and
unsure of himself. He never spoke up in meetings
anymore. Finally he decided he would never fit
in here, and requested a transfer to another office.

With Ralph gone, the Jonathan Livingston
Seagull Six developed a product that was creative and beautiful—but so impractical that no
one bought it. They could have used stodgy old
Ralph to bring them down to earth.

The problem with the Jonathan Livingston Seagull
Six was that they regarded Ralph's differences as
shortcomings. Instead of accepting the contributions
he could make, they regarded his practicality as a
fault and ostracized him until he felt he had no option
but to request a transfer.

Building a properly balanced team can be difficult.
Employees and managers alike may resist someone
who is "different," even if that person has comple-
mentary skills and attributes needed for getting the
job done even better. Differences in work style or personality can set off a series of prejudices in even the
most enlightened group.

Building trust and respect among your own work
group requires focusing on people's unique compe-
tencies. Different employees have the strengths to respond to different situations; it's not a matter of certain
traits being "better," but of which traits meet the needs
of the situation.

One way to help team members see and appreciate
those differences is through the use of instruments—

written tests or computer exercises that measure work styles. These non-judgmental tests not only determine an individual's work style, but show how the various styles complement each other. Instruments can help employees see that the sum of their differences is actually the work group's strength.[1]

Building a successful team requires instilling respect for differences and balancing strengths:

the need for action	vs.	careful analysis
big-picture thinking	vs.	attention to detail
willingness to take risks	vs.	caution and patience
a firm stance on issues	vs.	willingness to compromise
creativity	vs.	a sense of tradition

IDENTIFYING GROUP NORMS

Norms, or established ways of doing things, are developed on the job. In every organization there are certain *required activities* that people must do to perform their jobs. In performing these required activities, there are normally *required interactions* with other people. During these interactions, people develop sentiments or attitudes toward one another.

If the sentiments are negative, people will only interact when required by the job. But if the sentiments

107

are positive, people will plan other activities, such as going to lunch or socializing, so they can interact more. As they continue to develop positive attitudes toward each other, norms begin to develop. People want to be "part of the gang," so they don't go against the group's way of doing things.

Many managers fear group norms; they think any norms that arise out of socialization will work against accomplishment of organizational goals. That's not always true. As shown by the classic Hawthorne experiments, when group norms are directed toward hard work and quality they can be a very powerful influence on performance and human satisfaction.[2]

THE DIFFERENCE BETWEEN NORMS AND BELIEFS

Many organizations confuse norms with beliefs. They make no attempt to change the work force's norms, because they view such attempts as intrusions into employees' privacy. Instead, they impose rules: scheduled coffee and lunch breaks, no drinking at lunch. Yet these rules may run counter to the norms held by the majority of the work force. If a strong norm exists that "it's okay to take an extra fifteen minutes for lunch," employees will break the rules.

Norms, though, differ from beliefs. While beliefs are privately held thoughts that the individual feels very strongly about, norms are the aggregate of behavior that's accepted by the group. Norms can be changed. If the prevailing norms are endangering health, productivity, or safety, it's the company's responsibility to change them. If people find they will be or are being rewarded for adhering to positive norms, they will gradually change their behavior; but the first step in changing norms is to identify them.

Productivity norms shape people's attitudes toward their jobs. If these norms work against productivity by discouraging diligence and hard work, even an enthusiastic, highly motivated employee is unlikely to continue working hard.

Productivity Norms

It's normal around here for people

- To jump from crisis to crisis without any appreciation for long-range planning, priority setting, or time management.
- To complain about problems rather than doing something.
- To receive feedback only when they've "screwed up" rather than receive praise for a job well done.
- To reject new ideas out of hand: "We've never done it that way," or "I tried that once and it didn't work."
- To not be consulted regarding proposed changes that affect them.
- To be more concerned with the time clock than the quality of their work.

In the same way, most members of a group find it easier to conform to the prevailing health norms than to initiate behavior changes that will make them "stand out" from their peers. These dominant feelings of the group provide powerful deterrents to change. Health norms affect attitudes toward nutrition, stress management, and fitness, to name just a few. Let's take a look at a few widely held norms that can undermine wellness at work.

Nutrition Norms

It's normal around here for people

- To celebrate birthdays, retirement, holidays, and promotions with cakes and other sugary sweets.
- To grab a sweet roll and coffee for breakfast rather than eat a balanced, nutritious meal.
- To drink coffee or caffeinated cola drinks all day long.
- To have one or two drinks with lunch or to look for a drinking buddy after work.
- To gorge themselves on food whenever the company is buying.

Stress Management Norms

It's normal around here for people

- To avoid expressing feelings of conflict, to bottle them up and hope they'll just go away.
- To think you are "goofing off" if you take time to center and balance yourself (after all, the more frantic you appear, the harder you must be working).

Environment

- To overwork, often at the expense of their health.
- To handle conflict situations with hostility, blame, judgment, and guilt.
- To dispense aspirin and other pain medications to fellow employees for stress-related conditions.

Fitness Norms

It's normal around here for people

- To park their cars as close to the work entrance as possible, rather than walk several blocks.
- To use the elevator rather than the stairs for several flights.
- To view those employees who exercise on the job as "health fanatics."
- To bet on sporting games, rather than participate in them.
- Not to regard physical fitness as an important aspect of job performance.

One problem with these negative norms is that they're often perpetuated with the best of intentions. One manager of our acquaintance encountered norms that she felt were undermining health and productivity in her department—yet, she didn't know how to change them without offending her people:

"Liza," a thirty-four-year-old supervisor of a group of twenty salespeople, is personally committed to health. She runs, eats well, and manages her stress. She also prides herself on being a good manager. She sets realistic goals, motivates her people, and rewards

111

good performance. Her employees are unanimous in their praise for her management style. Yet Liza is having trouble with the health norms in her group.

Audrey, a dynamic salesperson, is having marital difficulties as she tries to blend two families. She suffers from frequent migraines and wild mood swings that make her difficult to work with.

Keith, the newest member of the group, is fighting hard for acceptance. Every Friday, to celebrate reaching the group sales quota he brings a cake to the office. When Liza arrives there's a piece already on her desk, tempting her away from her careful eating habits.

Liza knows that Keith's intentions are good, but he's making it hard on the three overweight people in the department who have been struggling to lose weight for the past year. In fact, if they spent as much time discussing products as they do discussing diets, Liza is sure they'd be way over quota every month.

Liza sees the health problems in her department and feels frustrated.

While she realizes she can't forcibly change the health norms in her group, she can try to establish a healthier environment. In working with her, we pointed out that she was starting from a position that had many positive elements. Her personal interest in health and her people's approval of her management style were strong pluses. Her people were generally committed to working well and staying well—as was shown by the fact that her overweight salespeople were aware of and concerned about their weight problems.

BEGIN BY ACCENTING THE POSITIVE

Liza decided to channel the health and productivity energy in her group by identifying the positive norms that the group wanted to reinforce while also identifying the barriers—the negative norms—that stood in the way of improved productivity and health. We told her about a group brainstorming exercise entitled *Going Well* and *In the Way*.

Liza started by telling the group, "Doing the best job we can depends on a lot of things. Some of them have to do with the system, some with the ways in which we work together as a department, others with having the personal health and stamina to get the job done. I'd like to see us create the healthiest, most productive environment we can. Let's do a little brainstorming and find out how we're doing, and what we can do better. Remember, this doesn't involve any judgment. Just let the ideas flow."

For the next hour the group suggested answers to two questions: "*What's going well?*" and "*What's in the way?*" Liza guided the discussion to include the company's system, co-workers, job responsibilities, and health, writing people's answers down on a flip chart at the front of the room.

Soon the group's health and productivity norms began to surface. Several group members mentioned obstacles to their weight-loss efforts: the company's junk-food vending machines and, of course, the Friday-morning celebration cakes.

The group identified negative norms that were getting in the way of health and productivity, but more important, they highlighted some strong positives such as teamwork, a feeling of camaraderie, and the group's feelings of respect and trust for Liza. At the end of the meeting, Liza had her secretary type up a list of the norms and provide a copy for each group member.

SHAPING THE GROUP'S NORMS

"We're not done yet," she told them. "We've identified the norms we want to reinforce, but some of them are more important than others. In order to rate which norms are most important, I want you to ask yourself a series of questions about each one." And she handed them the following list:

1. How often is this true—never, rarely, occasionally, frequently, always, don't know?
2. In terms of importance, is this: not very important, somewhat important, extremely important, critical?
3. Which way are we heading—getting better, getting worse, about the same?

"Once we know which norms are important," Liza continued, "we can reinforce the positive norms and reshape the negative ones. Ask the following questions about the norms that we've identified as most important":

1. Which norms are positive—and how can we strengthen these?

2. Which of the negative norms could we influence most easily, which least easily (rank order)?

Then Liza and the group met to determine group and individual goals. First, they chose one negative health norm and one negative productivity norm to work on. The group rewrote the negative norms in a positive way so that they had two positive goals to work toward.

She also encouraged each member to set two personal goals, one for health and one for productivity, and to identify the negative norms that would sabotage their efforts.

In their weekly meetings, the group reviewed progress toward their goals. Audrey started using a relaxation tape to handle her stress, and she initiated counseling sessions with her husband. Since over half the group chose "weight loss" as a health goal, Keith realized that his habit of bringing cake to the office would sabotage the goal of the group to promote slimness.

MORE ABOUT THE PHYSICAL ENVIRONMENT

As we have indicated, physical health as well as people's mental health can be affected by the environment. Organized health-promotion programs in the workplace can create a positive environment where physical health is the norm. More information on health-promotion opportunities is provided in Part II.

For most workplaces, though, a "healthy environment" will mean one that is free from obstacles to wellness. Unfortunately, most work environments contain elements that can be hazardous to your health. Two major issues that deserve mention because of their prevalence on the job are smoking and drug and alcohol abuse. Let's look at how managers can deal with these pollutants of the work environment.

POLLUTING THE WORK ENVIRONMENT: THE ISSUE OF SMOKING

Smoking at work is one of the most divisive issues a manager may face. Non-smokers insist on their right to a smoke-free environment, and smokers just as vehemently argue their right to practice the habit. Since blue-collar and less-educated service workers are more likely to smoke than white-collar professionals (43–47 percent vs. less than 30 percent in one study),[3] smoking can become a symbolic struggle between employees and management as well.

The growing body of evidence confirming how cigarette smoking hurts productivity and health, plus the growing militance of non-smokers, makes it impossible for a manager to ignore this environmental issue. Yet, it's difficult to find a way to deal with the issue fairly without offending either smokers or non-smokers.

For years, most companies had no smoking policies at all. It was almost universally, if tacitly, agreed that the right of smokers to light up where they pleased outweighed the right of non-smokers to a smoke-free environment. But that pendulum has swung the other way, so that now non-smokers are increasingly insistent on their environmental rights. For instance, a New Jersey Bell Telephone employee sued the company, charging that her co-workers' smoke irritated her upper respiratory tract, lungs, nose, throat, and eyes. The company has been ordered to provide a separate work area for non-smokers.

On the other side of the country, a recent survey of Seattle-area managers indicated that 53 percent are already giving hiring preference to non-smoking applicants. Convinced that smokers are twice as likely to be absent as non-smokers, nearly 90 percent of the respondents indicated that they would give preference to non-smoking applicants.[4]

It is legal to screen job applicants on the grounds of smoking. Discrimination against smokers in hiring and on the job does not violate equal opportunity statutes, as long as it is not a pretext for discrimination on the basis of race, sex, national origin, or religion. The U.S. Supreme Court has left little doubt that job-related factors such as smoking are legal criteria for personnel decisions.

Refusing to hire smokers, though, doesn't necessarily solve the problem. Many valuable, experienced employees are smokers; the question is not whether to banish them from the workplace but how to handle the accompanying environmental problems.

When they learn these facts about smoking, health, and productivity, many managers decide to use their influence to establish a non-smoking norm in the

117

workplace. One way to do this is to help smoking employees kick the habit. If your company offers a smoking-cessation program, you can encourage your people to take part. Participating in the National SmokeOut, providing self-help materials and video-tapes, or getting assistance from your local cancer or lung societies can help support your people in their efforts to quit. Some of the ways you can encourage a non-smoking norm among your people are:

- When leading a meeting, remove ashtrays from the meeting room and say, "I'd appreciate it if you didn't smoke during this session. We'll have frequent breaks for those who need to smoke." Smokers are getting used to this type of request, and most will go along with you. You'll also get a few strong nods of agreement from non-smokers.

- Once a smoking employee decides to quit, you can provide positive reinforcement. Try to empathize with what the former smoker is going through; abandoning a habit that has seen him through some tough times is like losing a best friend. He's going to need support from you and from co-workers for a long time. If your support slacks off, remember that the time most ex-smokers return to cigarettes is from three to twelve months after they quit. If necessary, mark your calendar to remind yourself to give bi-weekly praisings to the former smoker.

- Be relentless about repeating the health facts of smoking. Remember that your opinion and your influence can be important. For instance, many smokers fear gaining weight when they quit; you can reassure them that even if weight gain occurs, most smokers who quit return to normal weight in six months.

- Encourage the ex-smoker (and everyone else) to substitute an exercise habit for the smoking habit.

But what about those smokers who can't or won't quit, in spite of a supportive atmosphere? Dealing with the smoker/non-smoker issue may require rearranging the physical environment so that both sides have the maximum amount of freedom available.

First of all, consider the physical layout of your workplace. Is it a large, open, well-ventilated area? If so, desktop battery-powered air filters may be all that's needed to keep the air quality good for non-smokers. But if smokers and non-smokers are paired in smaller rooms where the smell of smoke lingers even after cigarettes are extinguished, creating a clean environment requires more creativity.

Those smokers with private offices find that rank has its privileges; they can indulge in their habit in their own offices, without offending co-workers. If you are a smoker with a private office, though, consideration of non-smokers is still an issue whenever you walk into another's work area with a lighted cigarette. If you are a smoking manager in charge of a group of largely non-smoking employees, your habit is a part of your management style that may alienate some people.

Managing the smoking issue becomes a matter of dialogue between you and your people. If smoking is

119

offensive to your employees—and don't assume it isn't just because they haven't told you—it may require some changes in management policies. Establishing smoking and non-smoking areas in offices and cafeterias or banning smoking at meetings can help improve the environment for everyone.

DRUG AND ALCOHOL ABUSE

Another area where managers bear responsibility for "cleaning up the environment" is the issue of substance abuse. Confronting an employee with a suspected chemical dependency isn't easy. The symptoms of certain medical conditions are similar to those of substance abuse; rather than risk making a mistake, managers often just let the issue slide. That's just one of the reasons managers use for not confronting impaired employees. Here are some others:

- "I don't know if the company will back me up." Employees who are dismissed or disciplined on charges of substance abuse sometimes bring action against their employer, something every manager wants to avoid. Yet, *not* taking action can involve even more serious consequences. Two men whose wives were killed by a drunk driver won a wrongful death suit against Otis Engineering Corporation, the driver's employer. The driver's supervisor, knowing he was drunk, had sent him home early from his shift— and on the drive home, the accident occurred.[5] See the checklist on page 124 for defining your company's drug and alcohol policies.

120

Environment

- "I don't see any users around here." When you hear the words "drug addict," what image comes to mind? A sickly, glassy-eyed bum lying in a gutter? Or someone who looks very much like you, your family, or your children? Substance abuse is a widespread problem which crosses all socio-economic, racial, and sexual lines.

- "It's not really hurting anything." Upper management in your organization might not seem to take a hard line on substance abuse, but what would be their reaction if tomorrow's headlines read, "Giant Drug Bust at XYZ Company—Dozens of Employees Involved"? Labor arbitrators have ruled that dismissal or disciplinary action is justified when an employee's drug use causes a loss of public confidence in the company, threatens the safety or well-being of fellow workers or the public, or interferes with completion of duties. In addition, many users need to steal from the company or sell drugs on company premises in order to support their expensive habits. Either way, the company loses.

- "It'll make me look bad." Some bosses see exposing their people's drinking or drug problems as an admission of their own weaknesses as managers. Similarly, companies often don't like to admit that there might be a substance-abuse problem in their ranks, for fear it will tarnish the company image. Anonymous eyewitness programs can help managers who want to do something without involving their names and positions.

WHAT TO LOOK FOR

Diagnosing employee substance-abuse problems isn't a manager's job. It's not supposed to be. You're not a physician, nor are you a law enforcement officer. You are, however, responsible for your people's performance and their safety.

Employee performance may begin to slip for a number of reasons. Perhaps they don't have the training they need, or perhaps they have personal, financial, or medical problems that affect work performance. Perhaps they are not getting proper direction and support for their tasks. If you have investigated likely causes and found nothing, and suspect an alcohol or drug problem, it is your job to confront the employee. But first you have to know what to look for.

No list of symptoms is infallible. Many of these symptoms could come from a number of physical and mental conditions other than drug or alcohol abuse, and being overzealous about watching for employee drug use can turn managerial concern into a "witch hunt."

To remain watchful for substance abuse without arousing hysteria, supervisors must be trained to handle the impaired employee. This training involves three stages: knowledge of the different substances that can impair performance, how these substances affect performance, and how to confront an employee with a substance abuse problem.

The following symptoms are often the result of impairment due to alcohol or drugs:

Physical signs: Exhaustion, untidiness, blank stare, slurred speech, unsteady walk, changes in appearance after break

Mood: Constant depression or anxiety, irritability, suspicion, mood swings

Actions: Argumentative, excessive sense of self-importance, avoids talking with supervisor

Absenteeism: Frequent "emergency" absences, leaving work area more than necessary, unexplained disappearance from job, frequent requests to leave work early

Accidents: Takes needless risks, disregards safety of others, higher than average accident rate

Work patterns: Inconsistent work quality and productivity, mistakes and carelessness, lapses of memory, increased difficulty in handling complex tasks

Relationships with others: Overreacts to criticism, withdrawn, problems at home, borrows money from co-workers

Drug and Alcohol Policy Checklist

Peter Bensinger, administrator of the federal Drug Enforcement Administration from 1976 to 1981 and currently President of Bensinger Dupont and Associates, a Chicago-based consulting firm, has compiled a checklist for determining whether your company's drug and alcohol policy is well defined:

1. Do you have a clear written company policy on drug and alcohol abuse?
2. Has your company given written guidelines on this issue to management and supervisory personnel?
3. Have you provided employees and supervisors with health and safety information on drugs and alcohol and made known the impact that such use on and off the job can have on job performance?
4. Have supervisors been trained to recognize key performance indicators often associated with drug and alcohol abuse?
5. Do your supervisors and management personnel know what to do if they find an employee who may be unfit for duty, under the influence of drugs or alcohol, or found in possession of an illicit or prohibited substance?
6. If drugs are found on company property or on company assignment, are local law enforcement agencies and company security forces promptly notified, and are procedures in place to accomplish this effectively?
7. Is there a medical resource designated for an ex-

amination, including a urine test, for employees
suspected not to be fit for duty?
8. Is there an Employee Assistance Program as a resource to help employees, and are they familiar
with how to participate in such a program?
9. Has the company briefed the union, if applicable,
on its drug policy and advised council presidents
that safety at work and fitness for duty are standards that all employees will be expected to meet?
10. Is the company directing its message at the co-
workers as well as the drug or alcohol abuser?
11. Are contractors formally advised that personnel
hired by them working on company property or
on assignment for the company would be denied
access if such individuals violate company drug
and alcohol policy and would be referred to local
law enforcement if found to be in apparent vio-
lation of the law?

If you suspect that one of your people has a drug
or alcohol problem but are unsure whether to confront the employee, ask yourself:

- Has the problem interfered with duties?
- Is it hurting co-workers, customer relationships, or the company's reputation?
- Does the problem threaten company property,
  other employees, or the public?

If the answer to any of these questions is yes, the
next step is to meet with the employee.

CONFRONTING THE IMPAIRED EMPLOYEE

If your company is one of the growing number of organizations offering Employee Assistance Programs (EAP's), confronting an impaired employee is made much easier; the intervention can include referral to a company-sponsored counseling program. Whether your company offers an EAP or not, a set of guidelines can help ease the friction of the confrontation interview. Remember, the employee's performance is the issue in a confrontation.

- The tone of the interview is all-important: support coupled with firmness, and above all, confidentiality.
- Stick to the facts. Review the employee's specific record of poor performance, state your concern, and explain what will happen if the situation doesn't improve.
- Advise professional evaluation and counseling if possible.
- Remain businesslike, non-judgmental, and firm. This is no time for "give-and-take."
- Conclude the interview with only two possible outcomes: the employee agrees to seek help, or denies having a problem. If he refuses to get help, remind him that continued poor performance will leave you no choice but dismissal.

Although confidentiality is important, so is documenting all confrontations, including the date, the specific offer of help, and whether the employee accepted or refused. This will help if the employee is later dismissed and brings action against the company.

As difficult as confrontation can be, it can be made easier by practicing a sample "script." A confrontation interview might go something like this:

> SUPERVISOR: "There are two things I'd like to say to you. First, your job performance lately has been way below par. You've been absent ten times so far this year, and six have been on a Monday. Four times during the last month I haven't been able to find you when you were supposed to be at your desk. You've postponed the deadline on your latest project twice already. If your performance doesn't improve, I'll have to take action.
> "Second, I'd like to say that I know you can do better than this. Your job performance two years ago was outstanding, and frankly, when I look at you today I see a different person. You seem shaky, and have difficulty concentrating. You used to be so easy to get along with, now you're jumping down people's throats at the slightest comment.
> "I know that frequently when people have personal or medical problems it can affect their performance. If you do have a personal or medical problem, I think you should seek professional help for whatever it is that's causing your performance to suffer. The nature of the problem is none of my business.
> "You understand that I'm not ordering you to do anything; the choice is yours. But you're a valued employee, and I must tell you that if your performance doesn't improve immediately, I'll be forced to take action."

127

At this point, the employee might try to cloud the issue with emotion by becoming defensive:

> EMPLOYEE: "Boss, I never thought you'd pick on me like this. If you think I'm an alcoholic, just tell me. If you're going to fire me, go ahead. But I'm telling you that it's not true, and I won't take this kind of smear on my character."

Resist the temptation to respond emotionally; remain rational and cool:

> SUPERVISOR: "I didn't say you were an alcoholic; I said that you should get help for whatever is causing the slide in your work performance. I want you to get help in correcting your performance before I'm forced to discipline you."
> EMPLOYEE: "So you're going to fire me if I don't get help?"
> SUPERVISOR: "On the basis of your poor performance, I think we would be justified in firing you. But we want to give you one more chance; we're offering you help to save your job. If you don't want to accept that help, then the choice is yours."

Confronting an impaired employee is a difficult task for any manager. After an interview like this, you're likely to ask yourself, "What went wrong here? How did the situation deteriorate to this point—and is there anything I can do about it?"

Behavior among any group of people is strongly influenced by the norms of that group. If substance abuse is tolerated or even encouraged by the employee's peers, you're going to have a widespread problem. Identifying and changing norms is one way to fight back, right at the source of the problem.

A GOOD ENVIRONMENT DOESN'T MOTIVATE—IT REMOVES DE-MOTIVATION

We know an organization's environment affects the way employees think and feel about their jobs; however, this effect only extends to a certain point. Environmental factors can determine whether or not employees are happy, but a good environment won't make them work harder.

A supportive work environment is important because it permits people to focus on their jobs. But people are motivated primarily by two factors that are outside the sphere of the work environment: the challenges of their jobs and the recognition they receive for accomplishment.[6]

Although providing a supportive environment doesn't make people work harder, it does keep them from quitting. It minimizes the time and energy spent thinking or complaining about the kind of supervision they receive, the people they work with, company policies, job security, or other environmental issues that are the basis of job satisfaction. The productivity improvement that comes from environmental changes reflects the removal of these de-motivators.

So managing the environment belongs to the "money in the bank" concept of human relations: If

129

you don't have any money in the bank, when you are in trouble you can't draw it out. Like money in the bank, a good working environment will stand you in good stead during hard times.

Similarly, the relationship between the environment and your people's health parallels the relationship between the environment and productivity: *A healthy environment doesn't guarantee wellness—it prevents illness.*

As we have seen, even the healthiest environment is no guarantee of wellness. Health is strongly influenced by the norms of work and social groups. Building a fitness center doesn't mean that people will use it, but it does provide the opportunity for health improvement as well as the message that the company cares about employee fitness. Providing a healthy environment allows for health in much the same way that a supportive environment allows for productivity—by removing the organizational obstacles that stand in the way.

There's more to the environment than mutual trust and respect, individual uniqueness, and health/safety; but concentrating your energies on these areas will give you the greatest return. The process of identifying norms, confronting the tough issues, and identifying the environment you want isn't an overnight process. But it can be done if you put it high on your list of priorities.

Chapter 5
Recognition

> People do better when they get
> feedback on their performance
> and recognition for their
> progress.

Why do people often not work up to their capacity?
And how can managers motivate them to put forth
their best efforts? These two questions are central to
most managers' jobs, and they revolve around the
issue of *motivation*.

First, let's look at the facts: How hard do people
work, and why don't they work harder?

William James of Harvard has found in his studies of hourly employees that they *could* work consistently at up to 80 to 90 percent of their ability if they were highly motivated. But he also found that, in general, working at 20 to 30 percent of capacity was the bare minimum required to keep their jobs.[1] In other words, an employee could put in as little as 20 percent or as much as 90 percent of potential effort, depending on motivation. That 70 percent differential, the result of motivation, represents what has been called the "commitment gap."

The commitment gap was identified in a three-year in-depth study by the Public Agenda, a non-profit, non-partisan organization founded in 1975 by Cyrus Vance and Daniel Yankelovich to study major policy issues. The Public Agenda interviewed 1,345 American workers as part of an international project, "Jobs in the 1980s and 1990s."

Although 52 percent of those surveyed agreed with the statement "I have an inner need to do the very best job possible, regardless of pay," only 23 percent said they were performing up to their real capacity. Forty-four percent said they put in only the minimum amount of effort required to keep their jobs. Why the huge gap between the amount of commitment people say they feel and the commitment that is translated into real work? An overwhelming majority of the Public Agenda respondents agreed on the cause—75 percent said "management doesn't know how to motivate workers."

That's only partially right. It's true that management doesn't know how to motivate people, but that's because people can't be motivated by external forces. Motivation comes from within. "Motivating people" actually means finding out how to tap into their inner commitment by providing the recognition they want.

FIRST, DECIDE WHAT TO REWARD

Here's where the commitment gap begins: with the concept of rewards. Ask any manager whether good behavior or bad behavior should be rewarded, and what will be the answer? Good behavior, of course. Yet most organizations are structured to reward the "squeaking wheel"—the complaining employee or the union that threatens to strike. Too often, people who perform consistently well are ignored by the reward system.

The first step in harnessing motivation is figuring out what you really want to reward. Consider the following for starters:

Long-range planning. Organizations usually offer the most recognition to those whose work is most visible and has the biggest short-term impact. Yet what about those whose work contributes to the company's long-term growth and health? Taking a long view of things can help put today's crises in perspective and help avoid crises in the future.

Working steadily and quietly. Organizations sometimes confuse activity with action and reward those who seem the busiest. Yet when the smoke clears, they find that the "quiet" workers may have been doing the bulk of the work all along. One company that recognizes the importance of serenity is MacDonald's Corporation, whose headquarters in Oakbrook, Illinois, includes an "isolation room" where employees can retreat from the bustle of the office to "center themselves."

Simplifying. With so many forces working to complicate our lives, employees who can simplify procedures at work are worth their weight in gold. One consulting firm uses a computerized "fog index" to help keep their written communications simple. Scores on the "fog index" correspond roughly to grades in school; any memo that rates above a 10 (tenth-grade difficulty) on the fog index is sent back to its author for simplification.

Creativity. Who says things have to be run "by the book"? Even in the military, where tradition is a way of life, creativity is being rewarded in some interesting new ways:

At the end of November 1983, William H. Taft IV, Undersecretary of Defense for Manpower Logistics, instituted the Model Installation Program. This three-year project is designed to help commanders run their bases more efficiently, save money, and improve morale by doing away with unnecessary regulations.

Commanders receive no money for instituting the program, but if they come up with ways to save money the base gets to keep it. The guiding principle of the program is "Unless it's against the law, you can try it." The program is currently being tested in 1 percent of shore-based installations.

Fort Sill Army Base in Oklahoma, a typical test site,

reports significant results from the first four months of the program. Two examples:

- Fingerprinting recruits used to take half a day, busing them from the training area at one end of the base to the administrative area at the other end. Someone suggested that the base use a portable fingerprint kit that could be brought to the recruits instead, saving hours of the recruits' time.
- The Army has always required a government driver's license for light vehicles, sedans, and vans, a process that consumed an enormous amount of time and expense. Fort Sill decided to accept a valid civilian driver's license; the change was immediately made Army-wide. Ironically, the base had been advocating the change since 1949 but only now has the organizational structure that allows it.

Risk Taking. By punishing mistakes, companies penalize unsuccessful risk takers and discourage anyone else from trying. Instead, they need to train people to be responsible and reward responsibility. Encourage people to go after what is possible, not just what is safe.

Solving Problems. Instead of hiding or sidestepping problems, cultivate a willingness to confront them; don't view problems as signs of failure but rather as an opportunity to learn.

Recognition drives motivation and productivity. To increase motivation and productivity, companies must figure out what they really want to reward, devise a way to measure it, then provide recognition for those

135

who achieve it. Recognition is the key to shifting from an organizational structure that rewards the wrong things to one that taps into people's inner motivation.

THERE ARE NO LOSERS, ONLY POTENTIAL WINNERS

Judging performance and making comparisons are an integral part of any manager's job. You need to know who's improving and who needs help; you need the facts on who your top performers are so they can be rewarded accordingly. But often these comparisons begin to color our views and managers start to label people, if only unconsciously: Jane is a winner; John is an underachiever; Beth has an attitude problem.

Using recognition effectively begins with a willingness to drop words like "underachiever" and "attitude problem" from your vocabulary. Instead, think of everyone as a potential winner; some are closer to achieving their potential, but everyone is on the same path. And the manager's job is to help them move along that path as quickly as possible.

Unfortunately, traditional performance reviews at most organizations don't allow managers to rate everyone as a winner. Most performance reviews are based on a structure that has been drummed into all of us since school days: the bell-shaped-curve rating system.

Many organizations use the concept of the bell-shaped curve in their employee rating systems. They use this logic: In a given sample of performers, the law of averages dictates that exactly half of them will be above average and half will be below, with most clustering somewhere in the middle.

Some rating systems allocate a fixed number of "points" in order to rank people more precisely. If you want to award more points to one employee to reflect improved performance, you have to take them away from someone else—even if the second employee's performance hasn't declined.

Under a system like this everyone knows exactly where they stand; if forty points is average for the group, they know what it means if they're ranked at sixty—or thirty—points.

But rating people as winners or losers is sure to have a negative effect on the morale, productivity, and health of at least half of the staff. Presumably the winners will be pleased with their ratings, but what about the losers? A supervisor in the electronics division of a large technology-based company summed up the effect of his company's rating system:

> Unfortunately, under the new Technical Performance Appraisal (TPA) system, we have to tell one-half of our engineers that they are below average. After we tell a man his score is below 40, he won't do anything for a month. He stews over his low rating, and may even take a few days' sick leave, even though he's not physically sick. After a month or two, we may be able to get him working again with the hope that he'll do better next year, but that's really a false hope.

He won't get a better score next year, because
the men above him will still be above him next
year, even if he does improve a little.[2]

This supervisor sees firsthand the negative health
and productivity effect of rating an employee a "loser."
But his company's rating system gives him no choice.
He could protest against the system by giving all his
people exactly the same number of points—but *his*
manager would probably object and rate him a "loser."

Even under a less rigid system, managers are under
pressure to rank employees. What do you think would
happen if you rated all of your employees as "excellent" on their performance reviews? You would probably see two conflicting effects:

- Employees would be proud of their ratings and
  eager to live up to them.
- Your manager would demand to know why you
  were making a mockery of the performance review system. After all, the system isn't designed
  to handle that kind of rating.

This is not to say that performance rankings are all
bad. Formal rating systems may be the organization's
way of providing consequences for behavior. For those
employees who are ranked highly, the performance
appraisal provides positive feedback and can be the
vehicle for raises, promotions, and other rewards. The
problem is with those employees who are further from
achieving their potential. How do you keep performance appraisals from discouraging them? One way

is to be sure the appraisals are based on clear goals and behaviors which have been agreed upon in advance. This process will be described more fully in the "Knowledge" chapter.

On a day-to-day basis, however, every manager at every level has a much more powerful, more immediate means of affecting employee behavior. When used properly, this recognition tool can be used to build and reward winners at every stage of development. Even when you face external constraints such as freezes on raises, promotion bottlenecks or a lack of other traditional rewards, this tool can be a powerful motivating device. It's called feedback.

FEEDBACK: THE CORE OF RECOGNITION

When people hear the word "recognition," many think of ceremonies where bonuses and plaques are presented for a job well done. While this type of recognition can be a powerful motivator, it doesn't happen often enough to affect what people do daily. The type of recognition that inspires or discourages high performance is the feedback you give on an ongoing basis—the pat on the back, the scolding, or even the lack of response that follows your people's daily efforts.

Feedback becomes an especially important motivational tool if it is the first response that follows an action. Whatever happens six months or a year down the road isn't going to have as much of an effect as what happens immediately after performance.

While it's nice to pat people on the back, meaningful feedback doesn't always mean praise. Properly delivered reprimands can improve performance and actually increase motivation by redirecting employees onto the path to success. But meaningful feedback depends on the ability to know what's really going on. This means knowing who's responsible for what, and being able to find out how individual performance contributed to overall results.

When an organization is structured so that the financial data, sales figures, number of calls per day, or any other necessary data are available, feedback can be both specific and meaningful. When areas of accountability are clearly defined, you can more easily determine whether people are performing up to the specifications of their jobs.

While any sort of feedback is better than none at all, feedback that helps employees perform at top levels follows certain rules of thumb. Meaningful feedback is:

> *Specific.* It describes what people did or what the results were, rather than assessing their worth. It does, though, include how the behavior makes you feel—proud, satisfied, disappointed, etc.
>
> *Immediate.* Let people know right away how you feel about performance. When you wait several months to deliver feedback—for instance, during an annual performance review—the hearer is more likely to argue over the facts than listen to the feedback.

Something the hearer can change. Telling an employee "I never have liked working with a woman" isn't likely to help anyone, while saying "I'd like you to start coming in to work on time" can have a positive effect on behavior.

For the benefit of the receiver, not the giver. Managers who give out compliments when they are in a good mood and shout when they feel bad only confuse their people. In giving feedback, ask yourself, Who will this feedback benefit? Me, by getting something off my chest? Or the receiver, because it will help him or her to feel good and learn?

More often positive than negative. For every *one* time you have to criticize, try to find at least *four* opportunities to praise. Research is showing that it takes this 4-to-1 ratio to achieve a health-promoting organization or relationship.

Verifiable. Was it heard by the employee the same way you meant to say it? The simplest way to check this is to ask him to repeat it. This way you can be sure feedback is really hitting the mark.

NEVER UNDERESTIMATE THE POWER OF DOING NOTHING

Most people think feedback comes in two varieties: postive and negative, sometimes called "praise" and "reprimand." Each type has a specific use in encouraging or discouraging the targeted behavior.

But there's a third type of feedback, one that is more common than both positive and negative: a neutral response. Perhaps a neutral response shouldn't be called feedback, since the employee sees no reaction from the manager. Neutral responses "sneak up" on managers; in the daily bustle of activity, it's easy to forget to deliver praise or reprimands when due.

The term "neutral response" is misleading, since the consequences of this type of feedback are anything but neutral. When employees get neutral responses for their behavior, motivation is eroded until they eventually give up trying and lapse into poor performance. The only time neutral consequences fail to discourage good performance is in those rare cases when the job is truly self-actualizing—when the behavior itself is so satisfying that no outside recognition is necessary.

Most jobs contain many tasks that are not self-actualizing for employees. In these cases, it's the manager's job to provide the feedback that gives satisfaction. When a manager fails to provide any recognition for a task, the result may be the same as if employees had been punished: They won't repeat the behavior. Let's look at how this worked for "Sarah," a copywriter at an advertising agency:

Sarah was the most recently hired copywriter on the staff, and she was eager to make a good impression. She came in early to work, was careful not to take long lunches or coffee breaks, and often worked late in the evening. Her boss noticed her diligence and thought to himself, "Now that's a good worker."

Soon, though, Sarah's fellow writers began ribbing

her about her long hours. "Come on, take a break," they urged. "We're going to take a long lunch today; come along. You're just making us look bad by working so hard."

At this point, Sarah's hard work was getting two different responses. Her boss noticed and approved— but didn't tell her about it. Positive thoughts don't count unless they're expressed, so her impression from him was a neutral response; she thought he neither noticed nor cared about her work.

At the same time, she was getting a negative response from her co-workers. She wanted to fit in, but they thought she was making them look bad. The negative response outweighed the neutral response, and Sarah began coming in late, taking long lunches, and leaving early.

Her boss noticed the change and called her aside. "Sarah, I've noticed that lately you've been slacking off. Now I expect a full day's work from you. If I catch you coming back late from lunch or leaving early again, we may decide that we can do without you."

Now Sarah is caught in a no-win situation. If she works hard, her co-workers are displeased; if she doesn't, her manager yells at her.

This is a situation where poor management of consequences ruined things for everyone. Sarah's boss failed to communicate his positive feelings; his neutral response turned her enthusiasm to bitterness and disillusionment. But the story could have had a happy ending if her manager had paid more attention to consequences:

The head of the copywriting department notices

that the new writer is putting in long hours learning her way around. Pleased, he calls her into his office and says, "Sarah, I'm glad to see you putting so much effort into learning your job. We both know that most of the other writers don't work as hard as you do, but I just wanted you to know that your hard work is noticed and appreciated. I'm sure you'll have a bright future with the company."

Sarah's co-workers give her a hard time about working so hard, but she just tries to be pleasant without letting her work slide. She remembers the boss's words and keeps reminding herself that her career is more important than becoming "one of the gang" by goofing off.

In the second scenario, Sarah's boss provides the positive response she needs to overcome the negative response of her co-workers. Simply telling her that her work is appreciated makes the difference between productivity and resentment. In the process, he is reinforcing the norm that he wants to establish: he's rewarding hard work and dedication.

> Managing consequences doesn't have to be complicated. There are just two secrets to making "what comes after" work as well as "what comes before":
>
> 1. Tell people what you think of their behavior and how it makes you feel. They can't read your mind, so a positive or

> negative response left unvoiced comes
> across as neutral.
> 2. Understand the power of a neutral re-
> sponse. Positive responses feed the be-
> havior; negative responses kill it, but
> neutral responses starve it to death. Ei-
> ther rescue the behavior or put it out of
> its misery.

BEYOND FEEDBACK: WHAT ELSE DO EMPLOYEES WANT?

Although feedback is one of the most powerful and most readily available means of rewarding employees, it's not the only means. Other, more tangible forms of recognition can be used to spark employee motivation.

The following list of ten motivating factors has been used for over thirty-five years to determine what employees want—and what their managers *think* they want. Look at the list and rank the three most important to you; then go over the list again and choose those that you think are most important to your employees:

IMPORTANT TO YOU?		IMPORTANT TO EMPLOYEES?
_____	Good working conditions	_____
_____	Feeling "in" on things	_____
_____	Tactful disciplining	_____
_____	Full appreciation for work done	_____

145

IMPORTANT TO YOU?		IMPORTANT TO EMPLOYEES?
_____	Management loyalty to workers	_____
_____	Good wages	_____
_____	Promotion and growth with company	_____
_____	Sympathetic understanding of personal problems	_____
_____	Job security	_____
_____	Interesting work	_____

Are your lists identical? Probably not, if you're like most people. In administering this test to hundreds of managers, we've found that what *they* want and what they think their *people* want are entirely different. Those most often ranked at the top for managers are interesting work, full appreciation for work done, promotion and growth within the organization, and feeling "in" on things. The top employee motivators— as managers see it—are good wages, job security, and good working conditions.

But what happens when we ask employees to rank the list for themselves? They name full appreciation for work done, feeling "in" on things, and sympathetic understanding of personal problems as their strongest motivators. That sounds more like what managers say *they* want, not what they think employees want.

Perhaps managers are making motivation more

complicated than it needs to be. As this little exercise shows, they tend to think that motivation is somehow different for employees than it is for them. Yet, there is a core group of motivators that satisfy basic needs for belonging and recognition—needs shared by everyone. A good rule of thumb is *Your people want 85 percent of the same things you do.* That is, most of the things that motivate you will also motivate them.

These motivators satisfy the 85 percent of needs that everyone shares. But what about the other 15 percent—that area where each employee is unique?

RECOGNITION IS AN INDIVIDUAL MATTER

No two people crave the same kind of recognition. Motivation programs that provide the same rewards for everyone aren't as effective as they might be, because they fail to take this basic difference into account. While it's important to provide some form of recognition if you hope to get repeated good performance, it's just as important to provide the right kind of recognition—the kind that a particular employee craves.

So when managers want to know "How do I motivate my people?" they're really asking the wrong question. You can't motivate people; but you can find out what's naturally motivating to them, then use it to recognize good behavior.

"Ted" was one of the top twenty salespeople in the United States for his company that year, and they wanted to reward him. So headquarters sent him a letter telling him that the company was sending him and his wife to Bermuda for a week with the other top salespeople.

Ted called his boss at headquarters and asked, "How much will it cost the company to fly me and my wife from California to Bermuda and wine and dine us for a week?" "Probably somewhere between eight and ten thousand dollars," his boss replied.

"Good! Send me a check."

Naturally, Ted's response caused a furor. At company headquarters, they wondered how Ted could be so disloyal and greedy.

When Ted found out that people were upset, he phoned his boss. "I need to straighten something out. Is this trip to Bermuda part of my job? If it is, my wife and I will be glad to go with the other salespeople. But if it's a reward for my performance, I'd much rather have the money to take my whole family to Hawaii."

In this case, Ted's company didn't understand what motivated him. Their attempt to provide what they thought was motivation—a trip to Bermuda for two— wasn't Ted's idea of a reward; rather, he looked forward to time with the whole family.

Finding the individual 15 percent of motivation requires creativity. For instance, most reward systems operate under the assumption that more money is the best reward. Yet money is a complicated motivator. It can be an external measure of the employee's worth; it can confer prestige; it can provide opportunities for freedom. It means something different to virtually everyone. Rather than relying on money as a reward

148

Recognition

in every situation, let's look at some other possible motivators.

For instance, if your secretary has done an outstanding job this year, you could reward her with a pay raise. But some other rewards might prove to be more motivating:

> *Time.* If she has children, she may have more need of time than of money. One afternoon a week for running errands, catching up on home chores, or just being with her children may be more valuable to her than a pay raise.

> *Ownership* in the company can be more satisfying than a raise. Your secretary may be willing to sacrifice short-term goals to help the company prosper, especially if she will benefit personally.

> *Challenging work* can be a powerful motivator, especially for those whose jobs are repetitive or boring. Seeing results of her work can make the process of the work itself more rewarding.

> Most employees will welcome the *opportunity for personal growth*. What skills would help your secretary do what she wants to do with her life? Give her the chance to take classes in management skills or accounting, whatever she wants to learn.

> Finally, don't let your secretary be an unsung heroine. *A recognition dinner*, plaque, or prize—especially when presented in front of colleagues, family, and friends—is a good way of letting everyone know that she is esteemed and valued.

These are suggestions for rewarding a hypothetical employee; you can use them as a starting point for creative motivation among your own staff. Rewards can be as diverse as membership in a club or the opportunity to work with a different group—as long as the reward answers the inner motivation of the employee.

How can you tell what motivates people? One way is to ask them, although sometimes employees are reluctant to tell you what they really want. If you can't find out by asking, observe what they do with their free time. Learn something about their lives, their families, their hobbies, their dreams. These will undoubtedly hold clues to their motivations.

RECOGNIZE STEADY PROGRESS, NOT JUST EXTRAORDINARY ACHIEVEMENT

Rewards and recognition are usually awarded at the end of a successful project. The members of the winning team are showered with awards, bonuses, and other reminders that their work is valued. This is appropriate; success should be recognized.

But what about cases where success isn't as evident? Your secretary may type perfect letters day in and day out without getting any recognition for her consistent good performance. Or a certain project may fall short of its overall goal—yet individuals who worked on the project performed very well. We tend to give the most recognition for visible success, yet we are surrounded by "hidden" individual triumphs every day.

Recognition

Rewarding individual progress requires the ability to judge people according to their own potential performance range, not yours or anyone else's. When you measure people according to the goals and standards of the top achievers in the department, you may not recognize how extraordinary their effort is *for them*.

For instance, we witnessed the following scenario in a Wall Street brokerage firm. A highly successful, internally motivated boss tried to sell his overweight secretary on the benefits of jogging.

"You'll just love it," he told her, and she found it easy to believe. She had heard her boss and his running cronies swapping stories and comparing times, and she had seen the obvious joy he got from running.

She began jogging with the idea that it would be fun. Three weeks passed and she had diligently run four times a week—and she still didn't "love it." In comparison with the other runners on the track, she felt dumpy and slow. She was sure none of them felt as sore and stiff in the morning as she did.

In fact, it would take many weeks for her to establish enough of a cardiovascular endurance level to even begin to enjoy jogging. In the meantime, she got little recognition for what she perceived as a heroic effort. Her boss asked her about her speed and encouraged her to enter races. When she told him that she didn't keep track of her speed and wasn't interested in racing, he was incredulous: "Then why are you running? How will you know how you're doing? Do you want to be just a jogger all your life?"

As a matter of fact she had been pretty proud of being a "jogger," but she was embarrassed to admit it to her boss. He seemed to expect so much more. Feeling resentful at his controlling behavior and a little ashamed of herself for "failing," she gave up her jogging program altogether.

What went wrong here? She started out full of enthusiasm and high hopes, but both quickly melted away in the face of what she saw as her "failures." Her boss's idea of success was to perform well in races, but she needed recognition and encouragement at her own level. She began to feel ashamed of her slow progress, and her boss's inflated expectations only made her feel worse.

Instead, she needed to be treated like a winner for the effort she was putting forth. If her boss had said, "I'm so proud of you for sticking with this program. I know it must be hard for you, but you're doing a terrific job," she would begin to internalize the feeling of winning.

With the proper encouragement, she would have been able to stick with the program until she actually began to enjoy the process of jogging. As the physical benefits of exercise began to outweigh the discomfort, her focus would shift from trying to please the boss to jogging because it was good for her. This example shows the importance of almost constant recognition, encouragement, and enthusiasm from everyone who hopes to help a "potential winner" internalize the motivation it takes to become a "proven winner."

Potential winners need an environment of trust and flexibility that allows them to take risks and move toward challenging goals. They need an atmosphere that encourages creativity and rewards achievement with task-specific, timely feedback. Most of all, they need to be rewarded for their efforts, not just their achievements.

RECOGNITION WORKS TWO WAYS

By now you know that recognition in the form of feedback or praise is one of the most powerful means of influencing behaviors. Two organizations we've worked with have taken praise seriously enough to build a formal program around it. A division of Holiday Inn has instituted a "praising coupon program" which provides guests with a booklet of praising coupons. The guests are instructed to look for "employees doing something right" and present the employee with a coupon filled out with a description of the positive deed. These coupons are then turned in to the manager, who praises the employee. The positive response to this program has affected absenteeism, morale, and turnover. As an interesting side benefit, guests have filled out three times as many hotel evaluation cards and over 80% have been positive—compared to before the praising coupon program when the majority were negative. After all, now the guest's job description was to look for positives!

A large textile company in the southeast, Milliken, conducts "bragging sessions" at its semi-annual managers' meetings. Emphasis is on sharing the successes Milliken is having in its quality and customer-service programs. Recognizing employee efforts is one of the surest ways to increase the likelihood of subsequent good performance.

The question is often asked: "What can I do about a boss who never praises anything I/we do?"

This situation is so common that it might help to reverse the way we think about it. When was the last time you caught your boss doing something right and followed up with a praising?

Managers, like employees, thrive on feedback and recognition. Praising—or even reprimanding—your boss shows that you *do* pay attention to his or her actions, and that you care about the way you are managed.

Just thinking about giving recognition to the boss is a little foreign to many of us. If we think about it at all, we tend to think that our bosses are being rewarded from above; if their jobs are tough, then they're being paid for it, aren't they? But if you think that managers don't need feedback, then imagine how *you* would feel if you received a sincere compliment on your management style from a subordinate.

Providing feedback to the boss isn't easy at first; after all, it's something most people haven't attempted before. Yet it follows the same rules as providing feedback to employees. For instance, task-specific, non-judgmental feedback is a much more powerful way to shape behavior than flattery or empty phrases. Consider a young man who is trying to establish a better relationship with his manager through feedback. Which example would be more effective?

154

> EXAMPLE A "You know, I'm really glad I work for
> you. Every night I tell my wife what a terrific
> manager you are. By the way, that's a lovely
> dress you're wearing."
> EXAMPLE B "Ms. Cleaver, that summary you wrote
> of last Friday's staff meeting was really help-
> ful. I passed it around to all my people, and it
> cleared up some questions they had."

Example A will put Ms. Cleaver instantly on the defensive. "What does this guy want from me?" she'll think. Blanket flattery—especially with personal remarks thrown in—sounds insincere and manipulative.

Example B, on the other hand, will give her a warm feeling. "What a perceptive young man!" she'll think. "I'm glad someone appreciates the effort I put into running this department!"

TRY IT—IT'S EASY

Perhaps you think we're putting too much emphasis on recognition as a shaper of behavior. Does a "pat on the back" really make that much difference?

Of course it does—as long as you keep the following in mind. First, it must be *specific* and *sincere.* Employees know when they're being "snowed," and they'll immediately wonder what's up. Sincere, specific praise for an effort, though, is always welcome. Second, recognition and praise must be delivered *often* if they are to be effective in changing behavior. "Proven winners" may be able to wait until performance review time to hear how they're doing, but most employees need frequent feedback along the way. Remember, it's almost impossible to give too much praise—and all too easy to give too little.

Chapter 6
Knowledge

> People do better when they
> know where to go, how to get
> there and why they're going.

ANTE UP BEFORE YOU GET IN THE GAME

Providing the knowledge people need to perform their jobs is like "anteing up" in a poker game. You have to put something of value into the "pot" before you can even get in the game. Knowledge about the job is a prerequisite for success.

In practical terms, that means making sure that you

157

and the people who report to you have an accurate picture of:

- what's expected of them
- what the goals are
- what a good job looks like
- how performance will be evaluated

Without anteing up the time and effort of providing knowledge about the job, you can't expect good performance from people. This is the "micro" aspect of knowledge, the nitty-gritty information about what you expect from people on a daily basis.

At the same time, you as a manager have access to more "big-picture" information about the company than your people do. You know the figures, philosophy, and organizational goals that relate to the importance or purpose of the jobs of your subordinates. It is the constant job of a manager to reinterpret for people the meaning of their jobs and how what they do contributes to the organization. One way to do this is to ask yourself, If this person's job were eliminated, what would go wrong or not get done?

This sense of being part of a larger purpose is widespread in smaller entrepreneurial companies, where everyone knows how their efforts contribute to the company's goals. Even the secretaries know that if they don't type that letter properly or answer that telephone effectively, a big project might be lost. In larger companies, though, people often lose sight of why they are working.

Knowledge

IT'S IMPOSSIBLE TO GIVE OUT TOO MUCH KNOWLEDGE

Managers hoard knowledge for various reasons. Perhaps they believe that "knowledge is power," so if they withhold knowledge from employees they maintain power for themselves. Or perhaps they don't want to overload employees with too much information, so they spoon-feed to people just the data they think are necessary to do the job.

Yet in most *successful* companies, particularly those in rapidly growing high-tech industries, employees have access to more information than they seem to need. In fact, sometimes people at these companies complain about having to go to too many meetings or read too many reports. Yet in a rapidly changing field it may be only a matter of days or weeks until this "unnecessary" information becomes a vital part of people's jobs. Making as much information as possible available to as many people as possible helps stimulate creativity and prevent mistakes.

As consultants, we've noticed that many companies seem more creative and vibrant when they are first starting out. For instance, one fledgling computer software firm had an area affectionately called "the ghetto," where programmers worked together in a large open space because the company couldn't afford dividers. Programmers occasionally complained about the lack of privacy, yet they all appreciated the free flow of information and ideas in their open work space.

The "ghetto dwellers" had a special sense of cama-raderie; they felt as if they were the real core of the company. Eventually the company installed fancy di-viders that separated the ghetto into rows of cubicles, and within six months all the original "ghettoites" had left for other jobs.

We've seen it time and time again: Once shoestring operations move out of the garage or warehouse and into a "real" office, they find that their informal lines of communication have been disrupted and they now need a "communications seminar" to teach people how to talk to each other again.

Essentially, we believe that it's impossible to hand out too much knowledge. There are some valid areas of secrecy within a company—for instance, salary in-formation—but for the most part, the more people know, the better. When employees know where the company has been, where it is going, and how they fit into that plan, they are more likely to be enthu-siastic about their jobs.

Although knowledge is vital at all times, it becomes doubly important when you are either planning a change or reacting to changes that have been thrust upon you. Now is the time that people want to know what's going on and how it will affect them; unfor-tunately, it's also the time that management tends to shut off communication.

Let's look at how sharing knowledge can help al-leviate the stages of concern and stress people go through during a change effort.

FOUR "LEVELS OF CONCERN"

The essential skill in sharing knowledge is knowing how much of it and what kind to share—and when. The need for knowledge corresponds to four predictable "levels of concern."[1] You can provide the information that can meet your people's needs for knowledge at each step along the way:

1. *Tell Me More.* This is simply an attempt to gain more information. At this point, people just want to know what's going on; they're hungry for the facts: who, what, when, where, and why. The most commonly heard phrases at this stage are "Fill me in. Brief me. Bring me up to speed." You can minimize stress and prevent harmful rumors from developing by filling in the details and providing your interpretation of the reasons behind the change.

2. *How Will It Affect Me?* After they find out what's going on, people will begin to personalize the event. They weigh the outcome versus the status quo. How will it affect their daily lives, their relationships with peers, the way they will be evaluated? At this stage they develop an emotional perspective; they begin to think of the change as either good or bad. They also begin to consider negative possibilities: "What if it doesn't work out?" Knowing how they will be affected personally can act as a buffer against stress.

3. *How Do I Get Started?* After they deal with the emotional side of the change, your people will need the step-by-step directions that clarify what you want

them to do. This chapter focuses on *how* to answer employees' needs during this phase by delivering change-related information clearly and concisely. The following chapter, "Style," discusses *when* to provide direction and support to help people adjust to a major change.

4. *What's The Benefit?* Most people resist change when they can't see any benefit to it. If the event will make their jobs easier, more rewarding, or more valuable, let them know. Even if there is no personal benefit, people will adjust more readily to a change when they can see its place in the "big picture." Don't just tell your people that these are "directives from above," but let them know what will be better after the change.

Now let's look at each of these levels in greater detail to see how you can provide the knowledge that helps people manage the daily changes which relate to their jobs. We'll also apply these steps to an organizational change effort—health promotion.

1. Tell Me More

Starting at the Finish Line

Imagine runners lined up for a footrace. The gun goes off, and everyone begins running. The problem is, no one knows where the finish line is. Is this a sprint or a marathon? Should they run as fast as they can or conserve their energy for later?

Some of the runners start out running as hard as they can, but soon they drop out exhausted. Others hang back, hoping they'll see the finish line in time for a final burst of speed. But most people clump together in the middle of the pack, thinking that everyone else must know what's going on, so it's safest to stick with the crowd.

Someone will win this race, but not necessarily because he or she is the best runner; the winner will have been lucky enough to set the right pace without knowing where the finish line was. The results of this race don't accurately reflect the abilities of the runners.

Working without clear goals and rewards is like running a race with an invisible finish line. Most employees opt for the safe route, doing just as much work as the next guy, but no more. Employees become confused and tentative, and the stress of not knowing what the goal is takes its toll on health and motivation.

How can you tell if your people are running toward an invisible finish line? One exercise we perform with managers and subordinates brings into focus just how much confusion exists about goals and priorities. First, we ask managers to list the five most important tasks their subordinates perform. Then we ask subordinates to list the five most important parts of their jobs.

Often the lists look as if they are describing two entirely different jobs! On the average, only 40 percent—two out of the five tasks—are the same for both lists. And when we ask both manager and subordinates to list the tasks in priority order, the confusion becomes even more apparent. What the manager thinks is important and what the employee thinks the manager thinks is important are two very different things. The result is wasted effort and resentment— and all of it could be avoided if employees and managers agreed "up front" what was expected.

GOAL SETTING: ESTABLISHING THE FINISH LINE

To clarify expectations, ask yourself: What is the goal of this job? Do my people know where the finish line is? This may seem simple, but conducting the above exercise at hundreds of worksites has shown us that clear goals and manager/employee communications can never be taken for granted.

One way to check your goal-setting process is to ask: Are your goals *smart?* Are they:

> *Stretching.* Stretching goals require a modest amount of effort—they are neither too easy nor too hard. By challenging our abilities they encourage personal growth and development and aid in skill building.[2]

> *Measurable.* This is probably the most important characteristic of a goal. If you say, "I want to improve our bottom line," how will you know when you've achieved the goal? Be specific, so you'll know how you're doing: "By January 1, I want net profit up from 10 percent to 12.5 percent." If you want "to improve customer service," a specific goal might be, "I want my people to be at work within ten minutes of starting time every day."

> *Agreed upon.* You may have a goal for your people to turn in their expense reports on time, but if they don't agree with that goal, how much

effort will they put into accomplishing it? An effective goal is agreed upon in writing, if possible, with measurements of progress and rewards for success built into the agreement.

Realistic. Can goals be accomplished in a set period of time with a moderate amount of effort? If one of your goals is to delegate to your new assistant the preparation of your department's budget within six months, you need to determine if it is realistic to expect someone who hasn't even gone through a full year's cycle to do all the planning and budgeting.

Trackable. Goals should be targeted toward specific behaviors, preferably ones that can be graphed or charted. If one of your goals for your people is to improve sales by ten percent, you might graph the number of telephone calls made by each salesperson every day. Watching the numbers on a graph rise provides positive feedback and makes the goal easier and more fun to achieve.

This method of setting goals can apply to any endeavor; see below, "Setting 'SMART' Goals for Health."

Setting "SMART" Goals for Health

Goal setting is just as important in making health changes as it is in managing a department. Health goals should also be SMART (that is, they should be Stretching, Measurable, Agreed upon, Realistic, and Trackable).

Setting "SMART" Goals for Health (*Continued*)

When setting health goals, keep in mind that long-term success depends on two factors: being able to see results in the short term, and having the support to maintain those results. For example, a long-term weight-loss plan started immediately before the Christmas holidays isn't likely to succeed, because celebrations with rich foods make short-term success unlikely.

Record keeping is the key to reinforcing long-term success. Missing a day or temporarily "backsliding" doesn't seem like an irreparable failure when you can look back through your records and see steady progress toward goals. Using graph paper or other logging systems, you can track minutes of aerobic activity, calories eaten, time spent practicing relaxation exercises, books read, pounds lifted, laps swum, or any other measure of progress toward goals.

---

2. How Will It Affect Me?

All changes involve emotions. If you ignore the feelings aroused by upheavals on the job, you may be ignoring potential problems. A little attention to the emotional aspect of what's going on can put people at ease and add to their confidence.

PINPOINTING THE FEELING

Managers often gloss over the emotions involved in a change because it's difficult to talk about feelings at the worksite. We've been taught that it's unprofessional. But in order to help your people perform at their peak, it may be necessary to deal with the emotions and fears that are part of any dynamic workplace.

"Reflecting" feelings is a skill that can help you understand how people react to a situation. As the name suggests, "reflecting" means focusing on the other person, demonstrating that you understand and appreciate what he or she is telling you. Reflecting feelings means paraphrasing what the other person is saying to you, with a special focus on the emotional component. For instance, imagine you are the supervisor of a new employee, a young woman who passes out soft drink samples in a neighborhood supermarket. She's been treated rudely by several customers and is on the verge of quitting. If she came to you with her complaints, you could reflect her feelings in the following way:

1. Focus on what she is feeling, and paraphrase that feeling: "You seem to feel frustrated and angry that people are being harsh with you when you don't deserve it."

2. Label and recognize her feelings. You supply these labels to show her that you understand what she is saying. If she labels the feelings herself, either ver-

bally or non-verbally, be sure to listen: "I understand that your frustration is causing you a lot of stress."

3. Describe the situation that causes the feeling: "Sounds like you feel hurt and helpless when customers snap at you or rebuff you."

4. Make sure your assessment is an accurate reflection of what she is feeling. If you are correct, fine; you can proceed to find a solution. If your assessment isn't quite accurate, then use that as a starting point to clarify things: "Am I in tune with you? Do you feel angry at the customers, or are you trying to tell me something else?"

This skill is an invaluable part of sharing knowledge. Denying your people's feelings by pretending they don't exist or aren't important can shut you out of an important source of feedback. When you "reflect" emotions to them, both you and your employees gain a greater understanding of their reactions to the situation.

3. How Do I Get Started?

At this point, employees want their manager to clarify exactly what they need to do to adjust to a new situation. If this step is skipped, it can result in anger, frustration, and wasted time and effort because people don't know how to approach the task at hand.

There are certain predictable situations when providing concise instructions is crucial. For instance, people are especially hungry for this type of knowledge when they are approaching an unfamiliar task

Knowledge

like learning to operate the department's new computer.

At times, providing the step-by-step knowledge will be unnecessary because people may already have the skills they need to adapt to the change. In this case it's inappropriate to provide more direction than the person needs. Normally, though, at least some of your people will need instruction in the day-to-day specifics of their jobs.

THE ART OF GIVING DIRECTIONS

Giving clear, concise directions is a skill in itself. Before you can show others what you want them to do, you must be able to:

- Pinpoint the result you want
- Express the goal clearly and concretely
- Make sure you are understood

When giving directions, it's especially important to state precisely what you want. A manager who issues a vague order like this one—"Karen, would you take care of that problem with Purchasing?"—is likely to get equally vague results. Be specific about what the problem is and what you want to have done about it:

"Karen, Purchasing has been taking up two weeks to get our orders out, and we can't wait that long. Contact Tom Smith in Purchasing and set up a meeting with him for tomorrow. Get the records for the last two months from the filing cabinet in Marie's office and show them to Tom so he'll know this is an ongoing problem. Tell him we need to get orders out within two days of the time Purchasing receives them."

If you want something to happen, tell people what you want to happen. The more concrete and specific you are in your directions, the more likely it is that things will get done.

CHECK FOR UNDERSTANDING AND AGREEMENT

Despite your efforts to be clear and precise, others may not hear the message you are sending. Or they may understand but have strong feelings that stand in the way of following your directions. If they understand an order but feel that it is silly or worthless, they aren't likely to carry it out effectively.

Lack of "present-moment thinking" may also prevent a job from being done. This happens when strong feelings—either pleasant or unpleasant—pull the listener's thoughts into the past or the future. While he appears to be listening to you, he is deeply absorbed in other thoughts and may not realize it until he attempts the task later and finds he cannot remember your instructions. Fatigue, illness, or external distractions can also stand in the way of clear communication. Often, when employees realize they misunderstood or didn't hear instructions, they are embarrassed to ask for clarification. They fill in the missing pieces with guesswork—and the job isn't done the way you wanted it.

For these reasons, part of giving directions is checking to make sure that your instructions are understood. Every time you give directions, check to make sure your people understand what you have said and know exactly what to expect. One way to do this is to ask them to paraphrase your directions:

"Now, give me your understanding of what I expect you to do."

"Would you repeat the steps I just outlined for this task?"

Sometimes, in addition to understanding, you want the hearer's reaction to your directions. In the last section, we demonstrated how to listen for emotions. If you suspect that employees might have objections, questions, or reactions, you can ask them:

"What are your reactions to this plan?"

"Will this plan cause problems for you?"

"What do you think about this?"

If you hear emotion-laden responses, use "reflecting" skills to figure out just what the employee is trying to tell you.

Giving directions requires more than just stating what you want. Sharing knowledge at this stage means making sure that your people:

- understand what the goal is and how they are to reach it
- have the means to reach it
- want to reach it

4. What's the Benefit?

After your subordinates have had a chance to learn what the change is about and what new behaviors are expected of them, their thoughts will turn to benefits or rewards. How will their lives or the organization be better now? If they feel that they will be better off personally after the change, they are more likely to support it.

At this point, most people will perform a personal cost-benefit analysis. Some will weigh the pros and cons of the change to decide whether they will support it. Others may not consciously go through this process, yet they still want to know "What's in it for me, my group, or the company?"

For employees in jobs that are intrinsically rewarding, a change may mean more responsibility or a chance for growth. For self-motivated individuals, it can be challenging. The new situation itself is the reward.

Other jobs, though, offer little intrinsic enjoyment. When employees feel that their jobs are unrewarding, change is often seen as a threat or a bother. For these employees, especially, it will be helpful to show how their participation makes a difference in the "big picture."

Every employee contributes something to the goals of the company. Yet, often employees don't have the company-wide knowledge that would allow them to see the importance of their jobs. As a manager, your role includes constantly reinterpreting your people's jobs to show their importance in the big picture. This type of knowledge can make everyday actions far more rewarding.

REMEMBER WHY YOU ARE GOING

This big-picture thinking is central to the return to basics evident in many industries. For instance, as hospitals face hard times, administrators remind themselves—and their employees—*why* they are in business.

Hospitals have a clear-cut mission: to heal the sick. But during the years when business was good and hospital beds were full, many hospitals began to unconsciously adopt the attitude that patients needed them more than they needed the patients. New technology and fancy equipment became more important than human relations, and patients began to complain that hospitals were "impersonal." Staff members who had joined the organization out of a commitment to personal values became disillusioned and cynical.

With rising health-care costs came pressures from insurance companies, government, and industry to cut hospital stays. Suddenly hospitals were faced with empty beds and increased competition for patients. Hospitals realized if they were to survive they would have to once again focus on *why* they were in business: to serve patients.

As a result, many hospitals are putting a renewed emphasis on "patient relations." Staff are reminded that what they do matters. A smiling orderly, a concerned nurse, a doctor who takes the time to answer questions can make the difference between empty beds and full ones. Columbus Hospital in Chicago offers a "money back" guarantee if patients are dissatisfied with the quality of care they receive.

Do your people know why they are working? In the daily rush to get things done, it's easy to lose sight of why their jobs are important.

If your group seems to have lost its "will to work," you can renew it by helping them see how their jobs contribute to the organization. Ask them to consider what would happen if their jobs were eliminated:

What would change in the department?

What wouldn't get done?

How would the organization suffer?

At the begining of this chapter we talked about the stress of running a race with no visible finish line. Imagine how much more stressful it would be if, in addition to not knowing *where* they were going, the racers don't know *why* they were running in the first place.

In the next two examples we'll examine how these principles apply both to a change in job function and to a larger organization effort: health promotion.

BEYOND BUSINESS AS USUAL

Business at Laura's restaurant had been falling off lately, so she hired a team of consultants to help her find out why. Together they pinpointed the problem: The waiters and waitresses had slipped into a careless attitude, and the customers had noticed.

So Laura, with the help of the consultants, set a goal of increasing sales. The way to achieve that goal, they decided, was to increase the number of comments unrelated to order taking that each waiter made to customers.

First, the consultants observed the staff at work, counting the number of non-food-oriented contacts between staff and customers. When they had a good baseline sample they made a chart showing average current contacts and the goal for each staff member, along with a timetable for achieving that goal. Another chart showed the baseline sales, with Laura's goals marked at the top.

Then Laura called a meeting to introduce the new system to her employees. The consultants explained the goal and the charts. In their consulting work with other restaurants, they told the staff, a program for increasing friendliness almost invariably resulted in higher sales and bigger tips for the friendlier waiters and waitresses.

To make sure the staff understood exactly what Laura expected, the consultants set up some role playing. All the staff agreed that it was more pleasant to be waited on by a smiling, outgoing waiter than by a stiff, wooden one.

For the next several weeks the consultants observed the staff and recorded the behavior changes and sales increases on the charts. Once a week, Laura privately gave each employee feedback on performance: "Your friendliness quotient took quite a jump last week, Jeff. Keep up the good work!" Or, "Debbie, you don't seem to be making much of an effort to be friendly. If there's something wrong, please talk to me about it. I know you can do a better job than you've been doing." When sales were up, she publicly congratulated the entire staff and reminded them that the efforts of each waiter or waitress were important.

One waiter, Don, didn't seem to be moving toward the goal that had been set for him. When Laura asked why, he told her that he felt uncomfortable "chit-chatting" with customers. So he and Laura agreed that he would become the restaurant's wine steward, a position where a more formal style would be a plus. He was happy in the new position, and wine sales at the restaurant increased.

After two months on the new program, Laura called another meeting to assess the results. Sales per customer were up, she reported, and the staff agreed that tips had improved. They decided to continue with the program until the staff felt secure in their new behavior, gradually shifting to a more informal program where they observed and encouraged each other, with senior staff instructing new employees in the friendliness system. Laura would continue to monitor sales and tips, conducting "refresher courses" in role playing when necessary.

This situation illustrates how one manager answered her staff's four concerns when she instituted a change in policy. By letting employees know what was going on at every step of the way, Laura ensured the success of her program. Let's look at how Laura's plan answered the four levels of concern:

Tell me more. Laura explained the goal to the staff, and why she felt it was important.

How will it affect me? Laura spelled out exactly what behavior change she expected: more contact with customers. The charts showed the waiters and waitresses their individual goals. In the process, she helped her staff deal with the emotional component of their jobs. For instance, when Don said he felt uncomfortable with his new role, she moved him into another job whose emotional requirements matched his personality. He was happier and his performance improved.

How do I get started? Role playing gave the staff a chance to learn their new tasks in a low-pressure situation. Employees were shown how to use the charts to track their progress.

What's the benefit? The waiters and waitresses were told that they could expect bigger tips as a result of their new behavior. Additionally, they shared in the excitement of watching the restaurant's sales skyrocket, knowing that their individual efforts were contributing to the success of the plan.

LARGER CONCERNS: KNOWLEDGE IN HEALTH-
PROMOTION PROGRAMS

The stages of concern are equally important in creating a successful health-promotion program. In the beginning, most participants want to know the scope of the program. Terms such as "wellness" or "health promotion" need to be defined. Orientation sessions must involve all levels of the organization. In larger companies, videotapes featuring key decision makers may be used for this purpose.

After receiving background information, people begin to wonder how the program will affect their lives. At this point the company's health policies must be clearly articulated, so employees can see how the program will affect them. For instance, on whose time do employees improve their health? Some companies provide education and screening during business hours, while others expect employees to participate in programs on their own time. Often, relegating health promotion to non-work hours sends the subtle message that health isn't really important to management, especially if other types of training programs have received company time off. An acceptable compromise might be to provide "matching" time. For instance, a one-hour seminar could be offered at four-thirty, for half an hour on company time and half on employee time.

Do programs include just employees or also their families? Again, separating health promotion from the

Knowledge

"rest of" your people's lives may make them think it's only a work-related issue.

Once the background information and health policies have been provided, concern shifts to more directive activity: How do I get started? This is the time to establish the benchmarks which will allow both individuals and the organization to measure progress. Among the tools which can make up a data collection system are:

- *Health Risk Appraisals* (HRA): These measure the risks posed by lifestyle choices and calculate the chances of becoming ill or dying from selected diseases. HRA's range from simple manually scored paper-and-pencil tests to interactive computer software that analyzes data instantly. The advantage of the HRA is that it is prospective. It identifies risks before they erupt into symptoms of disease, while lifestyle changes can still make a difference. The drawback is that they operate under the law of averages, which may or may not apply to a given individual. A number of these instruments are available commercially.

- *Lifestyle Analysis Questionnaire* (LAQ): Developed by Dee Edington at the University of Michigan, the LAQ takes the HRA one step further by considering other aspects of wellness. In addition to health habits, the LAQ measures job satisfaction, personality type, social support, stress, and other aspects of good health for a more complete wellness picture.

179

- *Fitness Tests*: These provide baseline measurements of physical conditioning which can easily be applied to large populations. YMCA's across the country have been using a test called "The Y's Way to Fitness" since 1973. Companies often contract with the YMCA to conduct on-site fitness testing of employees; one example of this is Union Pacific Railroad in Omaha, Nebraska. Herman Gohn, Project Director for Corporate Health Enhancement of the YMCA of the USA, says: "The tests are a motivator... especially the follow-up with the beginners who show significant results in the first ten to twelve weeks." See page 181 for details on what fitness tests include.

It is only at this point that people begin to internalize the personal benefits of a wellness program such as improved energy level, stamina, and endurance; greater self-esteem; better physical conditioning, and increased ability to manage stress. Unfortunately most programs get "sold" to employees on the basis of organizational benefits—such as decreased risk of heart attacks, lower medical utilization, or higher quality of life—*before* the individual concerns are addressed. The "sellers" then wonder why their well-intentioned efforts are met by indifference or sabotage. By understanding the issues behind the change process, you can address concerns with the type and amount of knowledge necessary for success. By "anteing up" knowledge, you let people know where to go, how to get there, and why they're going.

Fitness Testing

Baseline fitness testing usually includes the following:

- Skin-fold measurement with calipers, which provides a rough assessment of the percentage of fat deposited throughout the body. Underwater weighing measures body fat more accurately, but requires more sophisticated equipment.

- Cardiovascular endurance, including resting heart rate and heart rate recovery time, measured by using a bicycle or a step or a treadmill test to stress the cardiovascular and musculoskeletal system and provide an indication of the strength of the heart. In general, the fitter an individual, the lower the resting heart rate and the sooner the heart rate returns to normal following mild to moderate exertion.

    The most commonly administered fitness tests include a sub-maximal exercise test (the heart rate is not elevated to the maximum level) as opposed to the treadmill tests conducted with EKG monitoring, which push the heart rate to much higher levels.

    Other endurance tests include a two-mile run or a run for distance (how far a person can run in a given amount of time).

- Strength tests, which include a variety of sit-ups, push-ups, pull-ups, and bench presses to measure overall upper body strength.

- Blood pressure tests, which measure the systolic pressure (the initial surge of blood through the arteries) and the diastolic pressure (the resting level).

Fitness Testing (*continued*)

Normal ranges are less than 140 mm systolic and less than 90 diastolic.

- Flexibility, measured by toe touch or yardstick assessment in both standing and sitting positions. A normally flexible adult should be able to touch his toes without bending his knees.

- Additional tests may measure lung function, cholesterol (HDL and LDL) levels, and dietary habits.

Chapter 7
Style

> People do better when they are managed in a style appropriate to their level of skill and commitment.

If your people were robots, you'd have no trouble managing them. You'd have "conceptualizing" robots to do the long-range strategic thinking and planning, "detail" robots to take care of the day-to-day tasks, and "coordinating" robots to keep the other two types working together. Robots are always competent in their tasks, and always willing. They don't care about your management style; if you order or ask them, terrorize or plead with them, they still behave in the same way.

Humans, on the other hand, approach each task with a different measure of competence and commitment. Whereas any style will get the same results with your robot, effective management of people requires adapting your leadership style to fit the situation.

Although robots may be easier to manage than humans, they're not nearly as fun. While machines will perform consistently and readily, they are never creative. With a machine, you'll never get the unique synthesis of experience, personality and talent that makes management exciting. Ordering robots around is a static process; choosing the management style that will make the most of your people's unique abilities is dynamic and challenging.

This chapter provides a framework to help you select the appropriate leadership style for the various situations you will encounter on the job. These guidelines have evolved out of Ken Blanchard's approach to Situational Leadership,[1] which provides a means of assessing the task-related development levels of your people, and choosing a style that will help you get the best results. What Situational Leadership helps us see is that there is no "best" leadership style, rather "different strokes for different folks."

To be an effective Situational Leader you have to learn three skills: flexibility, diagnosis and contracting for leadership style.

FLEXIBILITY

Your leadership style describes how you behave when you are trying to influence someone else. For years it was thought that there were only two basic leadership styles—*autocratic* and *democratic*. In other words, you either told people what, when, where and how to do things and closely supervised them, or you involved them in decision making in a supportive manner. In recent years it has been found that there are four different leadership styles rather than two— each a different combination of directive behavior (autocratic) and supportive behavior (democratic). None of the four leadership styles is intrinsically "better" than the others; each is "best" when it matches the needs of the individual and the task. These leadership styles are:

- *Directing*. A manager using this style provides a lot of direction but little supportive behavior. Directing requires giving specific instructions on goals and performance, as well as close supervision and frequent task related feedback.

- *Coaching*. Direction is still high with this style, but now an increasing level of support begins to build a give-and-take relationship between employee and manager. When using this style, responsibility for the actual performance of the task begins to shift toward the employee as the manager asks for suggestions and ideas.

185

- *Supporting.* With this style, decisions around tasks are made jointly. The manager may ask, "What do you think we should do about the inventory problem?" Support remains high, but the managers turns over direction of the task on a day-to-day basis to the employee who has "learned the ropes." Frequent supportive feedback is an essential part of this style.

- *Delegating.* In this case there is little direction and little support. The manager may still identify problems, but turns task accomplishment and support over to the employee.

The four main leadership styles can best be illustrated with examples. A manager who has a new computer to install might handle the situation using any of the four styles:

- *Directing.* I want you to clear off that desk in the hall and unpack the crates. You'll need two extension cords and a small Phillips head screwdriver. John will show you step by step how to set the computer up; he's done that before. Call your wife and tell her you'll probably be home late for dinner.

- *Coaching.* A new computer just arrived. I thought you might set it up on that empty desk in the hall. I've set up a couple of them before, so I'll give you a hand to get it started. We'll need a Phillips head screwdriver and an extension cord. Can you think of anything else?

- *Supporting.* The new computer just came in. Take a look at it and get started setting it up. I appreciated the effort you put into the last installation. Is there anything I can do to help?

186

- *Delegating.* The new computer's in. Make sure it gets installed today.

While the goal to be accomplished remained constant throughout the four examples (installing a new computer), the leadership style (the amount of directive and supportive behavior) changed. A flexible manager is able to use all four of these styles.

FLEXIBILITY: NOT ALWAYS EASY

Encouraging growth by using the right management style requires a lot of flexibility. But just how easy is it for managers to be flexible? Most of us have one or two management styles with which we are most comfortable, and learning how to use a full repertoire of styles takes some effort.

Sometimes the nature of the job itself will call for a certain management style. The individual manager may not be flexible enough to assess the situation and use the right style. But if upper management can correctly pinpoint the problem, it might be easier in the short run to just hire someone whose style matches the job. Consider this example that was told to us by a consultant.

Before he began his consulting assignment, the top management of a company was faced with a plant that was in real trouble. They decided that what was needed was a new manager, so they transferred the manager from their top performing plant to the ailing plant. From a logical standpoint, if he could get good results from one plant, he should be able to get good results from another.

But it didn't quite work out that way. Not only did the situation not turn around, it got worse. So when the consultant began working with the company, the president mentioned over lunch:

"I've got a perfect example of the Peter Principle. I had a great plant manager and I sent him to this plant that was in trouble and he's fallen on his face. I guess he just rose to the level of his incompetence."

We, and this consultant, agree that managers often use the Peter Principle to get themselves off the hook. By blaming their people's incompetence, they avoid their responsibility for growing winners. We say to managers—and they don't like to hear it—that when they have to fire somebody, or find somewhere to "hide" them, or when they are worried about an employee's performance, they'd better get up to the mirror and take a look. In nine of ten cases, the biggest cause of the problem is looking right back at them.

So the consultant said to the president, "Don't fire this manager. Let me look into it." He found out that the new plant manager was a supporting and delegating manager in his two former management positions. Since most of his people had been highly competent, it was a beautiful match and he was very effective.

When he took over the new plant, everyone was expecting great things of him. It was a high visibility position, so he was under a lot of stress. Under stress in a new position, people play to their strengths. So he relied even more heavily on his supporting/delegating leadership style.

But the problem was that he was dealing with people who were not experienced or committed to the organization. For example, they were intent on building the size of their own departments regardless of whether they needed the people, just for their own power bases.

The plant manager involved the people who were causing the problems in decision making and they "ate him for lunch."

The consultant advised the president: "Get that plant manager out of there! Put him back in charge of an operation that's running well and he'll do a super job for you. Do you have anybody around here who eats nails for lunch?"

The president said, "Yes, but I was going to fire him."

"Why?" said the consultant.

"Because I had heard that autocratic leadership was no good," said the president.

"Forget that," said the consultant. "Send him into the troubled plant."

The president did just that—he sent his toughest manager to the ailing plant. The first few days he walked around the plant with a frown on his face, scaring everyone to death. The third day he called into his office one of his biggest department chairman and said, "Al, I want you to move your people into Room 110."

The department head said, "I can't do that, sir. That room only holds twelve people and there are thirty in my department."

"Right!" said the manager. "You've got till tomorrow to get rid of eighteen people."

So the new plant manager took command and shaped the place up. In a year and a half he had turned the operation around. But then the company had to replace him, too, because once it was running well he didn't have the flexibility to move through coaching, supporting and delegating styles.

So if you don't have flexibility, it is important as a manager to get yourself in a situation which lends itself to your favorite style(s). If that is not possible, then make sure you gather people around you who complement your style. That is, they can use the leadership styles that you are not comfortable using. A management team that duplicates your strengths will unfortunately compound your weaknesses.

Just because a manager is flexible, does that guarantee effectiveness? No! We both have worked with managers who were flexible, but who employed the wrong leadership in the wrong situation. They hounded the competent people, because they believed their results were crucial to their companies' success. But the inexperienced workers, who could have used more direction, were ignored because the managers did not think their performance made a difference. Thus they continued to flounder, while the experienced workers resented the managers' intrusion and their performance suffered as well. To be an effective manager, one must be able to diagnose a situation to determine the appropriate leadership style.

DIAGNOSIS

What are the key variables in determining what kind of leadership style a person needs? *Competence* and *commitment* determine whether or not people can and will do a good job. Competence and commit-

190

ment together make up a person's development level.

In determining competence, a manager asks him- or herself: Does the person have the knowledge and experience to accomplish the task with little externally provided direction?

In assessing commitment, the question is: Is the person motivated and confident about performing the job with little supervision?

It is important to assess developmental level by first taking a "snapshot in time," not an all-encompassing portrait. What are the individual's competence and commitment for this task or goal right now? This assessment shouldn't label anyone as "a competent person" or "completely unmotivated." Rather you are measuring where the person is on two scales, relative to a single task or goal.

Normally, as an individual learns how to perform a specific task, both competence and commitment levels increase. Progress from one development level to the next depends on a variety of factors: the difficulty of the task, the individual's unique talents, any outside help or hindrances that affect performance, and, of course, the direction/support he or she receives. Good performance is possible at any level of development as long as the appropriate leadership style is used; but the higher levels of development allow the best performance with the least supervision.

We see the natural progression of development most clearly in tasks that are challenging, that require a certain amount of perseverance and growth to master. Using golf as an example, let's track "Jeff" through predictable stages of development:

- *Enthusiastic Beginner*. Jeff decides to take up golf. He tells all his friends, buys clubs, and signs up for lessons. He is highly motivated, but not yet competent.

- *Disillusioned Learner*. After a few lessons, Jeff has just finished playing his first round of golf. "This is tougher than I thought. Maybe I don't want to learn this game after all," he thinks. His competence level has risen, but now he's not as motivated or committed. He's discouraged and considers quitting his golf lessons.

- *Reluctant Contributor*. Jeff sticks with his lessons and his game improves. He enjoys the game a lot—on the days he plays well. When he plays a good round, he really feels proud. But when he plays badly, he considers throwing away his clubs. He has grown quite competent, but his level of confidence and motivation fluctuates with his failures or successes.

- *Peak Performer*. After more lessons and lots of practice, Jeff is one of the top players at his club. He thoroughly enjoys the game, and feels that even when he doesn't play his best he can still learn something.

Development levels work the same way for the beginning employee, or the employee learning a new task, as they do for someone taking up a leisure activity. Just as Jeff almost gave up golf when he became a disillusioned learner, employees may become discouraged once the "honeymoon" of a new task is over. Gearing management style to employees' development level will help them over the rough spots and on to the next level.

Style

Let's look at how one employee demonstrates different levels of competence/commitment for her various tasks.

"Elaine" is a first-level manager for a computer software firm. She has always been considered a good employee, with considerable experience and a good attitude. When she or her boss set goals for the year and assessed her development level for each task, they found that her development level differed for each responsibility.

Responsibility	Competence	Commitment	Development Level
Write specifications for new product	High	High	Peak Performer. (She had been doing this task for three years.)
Write weekly status for staff	Moderate	Variable	Reluctant Contributor. (Her motivation varied according to how pressuring the rest of her workload was.)
Interview and hire three people	Some	Low	Disillusioned Learner. (She had interviewed several people but did not feel comfortable hiring.)
Give oral presentations	Low	High	Enthusiastic Beginner. (She was excited about the product and anxious to have it accepted, but had never spoken to a large group before.)

To be successful in all aspects of her job, Elaine clearly will need different mixes of direction and support provided externally by her manager or someone else. A question manager/employees need to ask themselves in this diagnostic process is: *For each area of responsibility* (task, goal, problem), *how much direction and support need to be provided by someone else, and how much can the person provide for him- or herself?*

MAKING A MATCH

The purpose in making this kind of diagnosis is to get a match between leadership style and the development level of follower(s). When that occurs the follower's attitude towards the manager is "I'm OK. You're OK." How do you know when that has occurred, or hasn't occurred? We have found it helpful to pay attention to nonverbal cues—watch people's eye contact, tone of voice and body position as they talk to you.

Any time you use a leadership style, there are three possible results (and corresponding non-verbal communication):

1. You achieve a *match*. Congratulations; you've chosen the right style for the employee's level of competence and commitment. In this situation, the employee will easily maintain eye contact. The tone of voice will be calm and relaxed and the body position will be square to you.

2. You *oversupervise* the employee. When you use a leadership style that is too directive for the employee's

194

development level, you'll notice the change in body language. Because over-supervision often leaves an employee with an "I'm OK. You're not OK" attitude, the eyes move away; they usually roll up as if the employee implores heaven to get this clown off his or her back. The body moves away, often sideward, and the voice becomes louder and more sarcastic. The general effect is like a teenager thinking, "How much longer am I going to have to put up with this?"

3. You undersupervise. When you choose a management style that provides too little direction for the development level, the employee will feel "Not OK" and less secure. The eyes will move down and inward and the body will follow, folding in and becoming small. The voice will be soft and tentative.

Of course, these "body language" interpretations aren't true 100 percent of the time. The point is that there are many ways to determine if the leadership style is effective, and sometimes nonverbal cues may be the first indication that something is wrong.

WHEN THERE IS A MISMATCH, EVERYONE LOSES

Choosing a leadership style isn't a mathematical formula. Your people's commitment and competence levels may be difficult to judge, or they may not fit exactly into the model. Circumstances may prevent you from providing the exact amount of direction or support that people need. It's OK to be "off" by a little; after all, situations change, and outside forces affect your ability to manage according to a set formula. But when managers stray too far from these principles, it can damage the working relationship.

The common result of both over- and undersupervision is that the employee has a negative view of the manager. In both cases, the employee feels that you are not doing your job; either you are not there to provide the direction he or she needs, or you are there all the time when you are not needed. The net result is a poor working relationship, low productivity, and unnecessary stress for both of you. Since we know that each of these results can have a powerful, negative effect on employee health, it becomes even more important to become a Situational Leader.

Now that you know about *flexibility*—the ability to use a variety of leadership styles—and *diagnosis*—when to use each of the four leadership styles—you are ready to learn the third and probably the most important skill—*contracting for leadership style*. We say most important because the contracting process in many ways integrates the five aspects of the PERKS system.

CONTRACTING FOR LEADERSHIP STYLE

There are several steps in contracting:

First of all, superior and subordinate have to *agree on goals and objectives*. That includes determining both key areas of accountability (responsibilities) and performance standards (what good behavior looks like). We suggest you and the people who report to you independently determine goals and objectives before you attempt to reach agreement. This will permit both of you to do your homework ahead of time. Then when you come together the first decision you have to make is who is going to talk first. If your subordinate speaks of his or her goals first, you as the manager should use good listening techniques. Make certain to paraphrase what your subordinate has said. Then a discussion and eventual agreement can take place more easily. If at any point during the discussion you cannot reach agreement on a particular goal, your viewpoint prevails.

Besides goal setting, this is a good time to agree on what incentives (hoped-for rewards) would be motivating to the subordinate. People need to know what's in it for them if they accomplish their goals.

Secondly, once you've agreed on goals and incentives, you and each of your people need to agree on their development level (competence and commitment) and the corresponding appropriate leadership style for each of their goals and objectives. Once again, we suggest that this analysis be done individually before you get together to reach agreement. This step requires that your people understand the principles of Situational Leadership. Remember, Situational Leadership is not something you do to your people but something you do *with* them.

To illustrate this step, let's look again at Elaine, our first line computer software manager who is at a different level of development for each of her four responsibilities. Once Elaine and her manager had agreed on her competence and commitment for each of her responsibilities, they then decided on how best her manager could help her in each area. The results of their discussions are given below.

Goal	Compe-tence	Commit-ment	Develop-ment Level	Manage-ment Style Required
Write specifica-tions	High	High	Peak Performer	Delegating
Status memos	Moderate	Variable	Reluctant Contribu-tor	Support-ing
Interview	Some	Low	Disillu-sioned Learner	Coaching
Oral pre-sentation	Low	High	Enthusias-tic Begin-ner	Directing

Elaine's manager agreed to adapt a "delegating" style for the job of writing specifications, setting a timeframe for completion and checking in with her occasionally, but for the most part leaving the project to her. With delegating, Elaine is now in charge of communications between the two of them. If she needs any help or has a problem, it is *her* responsibility to call her boss. When she does that the agreement would be for her boss to shift to a "supporting" style and listen to and support Elaine's efforts. If it turns out that the situation requires more direction, a movement back to "coaching" would be agreed upon. There would never be a need to move all the way back from "delegating" to "directing." That would set up the "leave alone–zap" approach that is very demotivating.

For writing status memos, Elaine and her boss agreed that a "supporting" style would work best since she has the skills to do the job but sometimes lacks confidence that she is doing it correctly. As a result, her boss essentially asks Elaine, "How would you like your 'strokes' (support and recognition) delivered? How about lunch every two weeks when you can show me what you have been doing in this responsibility area and I can listen, praise and support your efforts?" If Elaine and her boss agree on that strategy, they both write down in their calendars exactly when these "moral support" luncheons will occur.

When Elaine and her manager agree that she needs a "coaching" style for her interviewing responsibility, that means that her boss is in charge now. With delegating and supporting, Elaine was at the helm. Now Elaine's manager might say "Get out your calendar and let's agree to meet together on interviewing strategies and techniques at least twice a week for the next month. How about Wednesday and Friday from 1:30 P.M. to 3:30 P.M.?" If they agreed on those times, both of them would write those times in their calendars. At these meetings Elaine's manager would teach her the skills that will make her a more successful interviewer and provide feedback and encouragement to build up her enthusiasm and confidence for the task.

Finally, Elaine and her manager agreed on a "directing" leadership style for her oral presentation training. This might mean sending her to a concentrated public-speaking program or working closely with her boss. In either case, Elaine needs someone to show her a presentation format, listen to her practice, correct her mistakes and provide task specific feedback to help her improve her presentation skills.

We hope it is clear that this step of agreeing on development level and the corresponding leadership style is not a "loosey-goosey" process. You and your subordinate do not just agree on a particular leadership style and then leave it at that. You specify exactly what that means in terms of when and how often you will meet on a particular task and what your roles will be.

In the third and final step of contracting, you and your subordinate implement what you have agreed upon for each responsibility area. If you agreed that you were going to meet twice a week, you do exactly that. And in the process of keeping your agreements with each other, both of you continually look for performance indicators that suggest the manager shift toward increasing delegation.

HOW TO HANDLE REGRESSION

So far we've looked at how people progress from a beginning development level to peak performance. Unfortunately, the trend isn't always upwards; sometimes, for various reasons, people slip back in development level and management style must be adjusted accordingly.

There are a number of situations where employees regress in development level. If a top performer is passed over for a promotion, motivation may suffer, turning a former peak performer into a reluctant contributor. This is as far back as a top performer will slide. If he or she already has the competence to do a good job, it's impossible to go back to the enthusiastic beginner or disillusioned learner level for this task. In these cases any shift in leadership style should be in the direction of a supporting style. But if an employee develops drug or alcohol addiction or other serious personal problems, competence as well as motivation may decline. In cases like that, a manager will have to become more directive and even involve outside professional help.

Another cause of regression is failure to use the right management style in the right place. Some managers, anxious to be liked, are most comfortable with an extremely supportive style and use it before their people are ready. They call a group of eager people together and say, "These are the problems we face. How do you think we should handle them?" The employees don't yet have the experience to handle this style, but they have the enthusiasm to want to try. Of course, with this combination they are going to have more failures than successes. Soon their motivation levels will drop until they're a set of disillusioned learners—barely competent and now with low motivation—undersupervised in a delegating style.

Too often when employees regress, managers overreact. We see the same pattern in parents. A child has been behaving, cleaning his room and making his bed every day. But when he fails to get any positive feedback for his good behavior, and when he sees other children's bad behavior going unpunished, he regresses. His parents react by going immediately from a delegating style to directing: "Clean your room right now!" The child already knows that he is supposed to clean his room: what he needs is more support, not more direction.

One final cause of "regression" is a promotion or change in duties. We put "regression" in quotation marks because although performance may decline, it's because the employee's duties—and development level—have changed, but the management style hasn't. For instance, when a telephone operator is promoted to supervisor the general expectation is, "She was a great telephone operator, and now she is going to be a great telephone supervisor." Even though managers know that doing a job and supervising someone else's doing the job require two different sets of skills, they rarely take the trouble to reassess the newly promoted employee's development level and adjust their management style accordingly. Too few managers realize their own role in changing their leadership style backward to coaching or directing so they can give employees the proper direction and support they need to move competently into new roles.

> Using the right leadership style
> is managing for wellness.

Style

The health benefits to the employee of being managed according to the Situational Leadership principles are enormous. People will know that if they stretch themselves or get promoted into new goal areas, they can "contract" for the leadership style they will need to succeed in their new task. They will not be left to flounder and possibly fail. Similarly, in areas where employees can provide their own direction and support, they will be able to be "intrapreneurial" in their tasks and work autonomously. Given a proper match of leadership style to development level, employees can move closer and closer to high level wellness in their work.

CONTRACTING FOR LEADERSHIP STYLE AND THE PERKS

Contracting for leadership style beautifully integrates the five stages of the PERKS system. And the PERKS system helps people feel good about themselves and produce good results. Involving people in goal setting and determining how they should be supervised is the ultimate in *participation*. This process creates a trusting, caring *environment* where people think they can win. Determining what kind of *recognition* people would like to receive for good performance is built into the process. Since in contracting for leadership style there are no surprises, people have the *knowledge* necessary to facilitate good accomplishment. And finally, contracting is all about determining the appropriate leadership *style* to use

203

in what situation. Altogether, PERKS moves people to high levels of performance and self-esteem that greatly affect their overall sense of health and well-being.

PART II

*

Resources: Health-Promotion Programs

Chapter 8
Establishing Wellness Programs in Your Company

This "Resources" section is designed to help those who are interested in starting a health-promotion program at the worksite. We'll begin by providing some examples of what other organizations have done and are doing to keep their people well. As you'll see, there is a broad range of efforts which fall under the term "wellness program." After discussing several of these, we'll help you decide how best to focus your efforts.

AT&T COMMUNICATIONS: TOTAL LIFE CONCEPT

AT&T Communications' broad-based Total Life Concept (TLC) program is an excellent example of integrating health promotion into the organizational culture. The pilot phase of the program, which ran from May 1983 to May 1984, proved so successful in improving the risk-factor picture and employee attitudes that the company plans to expand the program throughout the AT&T Communications employee network by 1988.

TLC's comprehensive approach to healthy lifestyles is spelled out in a memo detailing the results of the pilot program: "TLC is specifically designed to reduce employee health risks and associated health care costs, improve employees' attitudes and reinforce the desired organizational culture."

According to Richard Bellingham, President of Possibilities Inc. (the consulting group that assisted in the design and implementation of the program), integrating culture change efforts with health promotion had additional payoffs.

> AT&T had just made a historic announcement that it had settled an 8-year-old antitrust suit with the U.S. Justice Department by agreeing to divest the Bell Operating Company, representing almost four-fifths of its total assets. Top management recognized the scope of the challenge it faced in breaking up the world's largest corporation, and the impact this would have on the

people who would have to do it and live through it while trying to operate a successful business at the same time.

In the health-promotion effort, we were able to help people appreciate on a personal level the same phases of change that the organization was experiencing. An integrated change effort prevents the predictable toll people have to pay when going through change without being equipped with the skills to manage it. Health promotion, in this context, becomes an excellent means of reinforcing the desired values of the corporation.

Employee participation in the pilot program consisted of three phrases. First, each participant completed a Health Risk Appraisal (HRA) in May 1983 that detailed both physical condition and attitudes toward health and work.

Then the company offered various lifestyle "modules" designed to provide employees with the knowledge, skills, and support they needed to successfully change health behavior. The modules ranged from exercise classes and smoking-cessation clinics to stress management and interpersonal skills. Finally, in May 1984 the participants completed another HRA to determine how their health and attitudes had changed over the course of the project.

While the pilot achieved substantial health improvement in a number of areas, analyzing the cost-benefit ratio of just one risk factor—cardiovascular disease—shows high promise that the program will be a financial success. Applying the risk reductions achieved in the pilot program to the company's 110,000 employees, Dorothea Johnson, M.D., AT&T Communications' Corporate Medical Director, expects the program to reduce heart attacks by 374 cases over the next ten years, saving the company an estimated $93 million.

Besides the dollar savings in risk reduction, analysts of the pilot program saw an improvement in employee-reported attitudes as well. Participants in the program reported an increase in the belief that AT&T Communications is interested in their welfare, an increase in their commitment to improved health behavior, more enthusiasm, and more positive attitudes toward work.[1]

HEALTH PROMOTION AND COST CONTAINMENT GO HAND IN HAND

As AT&T Communications found, a well-planned program at the worksite can not only improve employee health but may also help contain skyrocketing health costs. Many programs are aimed specifically at reducing costs, often by restructuring employee health benefits. There are a number of ways this can be done:

- *Changing co-payments* by raising the amount of the deductible paid by the employee or changing the proportion of insurance-paid to employee-paid expenses—for instance, from 90 percent insurance-paid, 10 percent employee-paid to 80 percent–20 percent.

- *Alternate health-care delivery systems*:
  Health Maintenance Organizations (HMO's) provide, for a set fee, all health care services (of-

fice visits, immunizations, health education) as well as traditional treatment of illness. HMO's emphasize preventive health-care practices and cost-effective treatment plans, using hospitalization only when necessary. In a controlled study of patients randomly assigned to either an HMO or a fee-for-service plan, hospitalization rates for those in the HMO were 40 percent less than the fee-for-service plan. The HMO group also incurred 25 percent less in health-care expenditures.[2]

Preferred Provider Organizations (PPO's) consist of a group of doctors with whom the organization contracts for various health-care procedures at preset prices. Employees can choose from any doctor in the PPO.

- *Utilization reviews.* Companies set up review boards to scrutinize all employee hospitalization claims for fairness of charges.

- *Coalitions.* More and more companies are joining business-oriented health-care coalitions like the Washington or Midwest Business Group on Health. These coalitions share health-care experiences and cost-cutting ideas.

Companies across the country have had great success with these cost-containment policies. Some pursue a single strategy, while others implement comprehensive plans that combine elements of several different approaches.

Health care costs at *Aluminum Company of America* (Alcoa) had increased by more than 250 percent since 1976. The solution, according to Richard G. Wardrop, General Manager of Compensation and Benefits, was to "put employees back in the picture of making some decisions around health care."

In negotiations with the Aluminum, Brick and Glass Workers (ABGW) union, Alcoa came up with a plan that increased employee-paid co-payments—but also provided each eligible employee with a reimbursement account containing $700 at the beginning of the plan year. As the employee accumulated medical expenses, co-payments were made from the account, not the employee's pocket. And best of all, any amount left in the account was handed over to the employee at the end of the year.

The plan, which went into effect at one plant on June 1, 1984, is expected to save Alcoa about $4 million in projected health-care costs in 1986.

In addition to the reimbursement plan, Alcoa sponsors community-wide health-education programs, fitness programs, and "fun runs" in several of its plant communities.

Westlake Community Hospital in suburban Chicago has resorted to "bribery" to encourage its employees to stay well. When costs for the hospital's traditional first-dollar-coverage medical plan were predicted to rise 38 percent from 1982 to 1983, hospital officials realized they needed to come up with some creative solutions to the problem of health-care costs.

Jane Garoppolo, Westlake's Manager of Wages and Benefits, studied the problem over a six-month period and came up with an assortment of plans that have reshaped the benefits policy at the hospital. Beginning January 1, 1984, Westlake instituted a Flexible Benefit

Account (FBA) for its 650 employees. Full-time employees receive $200 a year ($100 for part-time) in a non-taxable account to pay medical expenses not covered by the hospital's insurance plan. This money may be used for health-related programs, health insurance co-payments, physician charges in excess of usual and customary fees, and for non-covered dependents.

In addition, employees can get a "bonus" for living healthy lifestyles—an extra fifty dollars for each of three measures: maintaining normal blood pressure, remaining within a desired weight range, and not smoking. If an employee is involved in an accident while wearing a seat belt, the plan covers 100 percent of emergency room charges.

Another problem Westlake faced in 1982 was absenteeism; unscheduled absences averaged eight days per employee. So Westlake combined its former sick days, vacation days, and holidays into a thirty-day (thirty-five days for those with more than seven years' service) Paid Earned Time (PET) plan where employees can "bank" accrued time off and plan absences more carefully.

Westlake's creative answers to health-care problems have really paid off. Since PET and FBA were introduced, unscheduled absenteeism has declined by 15 percent and medical claims have risen 24 percent less than projected. The money Westlake saved on medical costs enabled them to launch their Aimwell program, which provides a series of health-education workshops covering wellness, fitness, lifestyle management, and an "independent study."

(The 1984 tax act may limit the creativity of the FBA plan, forcing employees to decide in advance how much to spend on which benefits, with no payout of the unused portion at the end of the year.)

Zenith Corporation's annual health-care bill per employee rose from $620 in 1977 to $1,192 in 1983. At that rate of growth, the company predicted, health care would cost $1,776 per employee by 1986.

To combat this rise, the company decided to take an aggressive approach to cost containment. The aim of the Zenith Medical Services Advisory Program is to cut costs and to turn employees into better health-care consumers through education. To do this, Zenith hired a medical services adviser to help salaried employees faced with hospitalization choose the most cost-effective health-care options. The counseling and mandatory second opinions for certain surgical procedures have cut Zenith's hospitalization costs (which represent over 60 percent of the company's total health-care bill) by $4,500 a week.

In addition, the company publishes a health newsletter that reinforces the informed-consumer message with informational articles and quizzes. The company also provides an alcohol and drug counseling and referral service for impaired employees.

BATTLING RISING COSTS WITH EDUCATION

Educating employees to be better health-care consumers means helping them learn the communications skills they need to be assertive and ask the right questions. The chart below illustrates some of these questions.

Asking the Right Health-Care Questions

Becoming a wise health-care consumer involves:

- Selecting the right doctor
- Learning to talk to health-care professionals
- Asking for generic drugs
- Learning the importance of a second opinion in eliminating unnecessary surgery
- Emphasizing self-care and preventive medicine
- Asking questions such as:
  Do I understand what's happening to me?
  Do I need these drugs?
  Do I need these tests?
  Do I need to be hospitalized?
  How can I prevent this in the future?
- Knowing how to care for simple illnesses at home[3]

The communications skills listed in the chart, often called medical self-care skills, were the subject of the Cooperative Health Education Program, a quasi-experimental study involving 2,358 households conducted by the Center for Consumer Health Education (Reston, Virginia) and two health-maintenance organizations, the Rhode Island Group Health Association (Providence) and Prime Health (Kansas City).

215

The experimental group, which received education on self-care, was able to decrease total health-care visits an average of 17 percent and minor illness visits 35 percent. Cost savings for the project were impressive. For every $1.00 spent on educational intervention, one group saved $3.43 and another saved $2.41.[4]

GETTING STARTED

Organizations that decide to sponsor wellness programs don't all begin from the same starting line. Each of them is at a certain stage in the development of a culture that supports health. Those that have more experience in the wellness field, such as the examples we have used, may already have benefited from seeing how PERKS can be integrated into their ongoing efforts.

For those organizations that are less advanced, the most pressing question is: How do we get started? Initiating a wellness program is similar to other organizational efforts: It proceeds along predictable lines, ranging from assessing needs to implementing the program. This section looks in greater detail at the preliminary steps leading up to a successful health-promotion program. While each organization's situation is unique, these general principles apply in most cases. Knowing the steps toward health promotion will help get you going in the right direction.

BUILD ON EXISTING EFFORTS

Many organizations already sponsor programs which may fall under the mantle of health promotion. Taking stock of existing programs can provide a base on which to build any new efforts. If your organization has been involved in health promotion, now or in the past, ask the following questions:

- Is (or was) the effort successful?
- Can it be expanded to include your new goals?
- What would the founders or administrators of the program do differently next time?

At some point, someone may resurrect any past failures and object to new efforts on the grounds that the old one didn't work: "Do you think this program is a good idea? After all, we tried that stop-smoking program a couple of years ago and it didn't get off the ground."

A past failure doesn't mean that health promotion is doomed to fail now. Perhaps employees weren't ready to take advantage of health promotion a year or two ago, but now they are more receptive to the idea. A previous failure may have been due to a flaw in the program's design, or to a negative influence in the work climate that doesn't exist anymore.

217

CONDUCT A NEEDS ANALYSIS

In the "Participation" chapter we cautioned, "Before you start, get the facts," and provided a questionnaire to help determine what employees want and need in health promotion. In addition to the questions in Chapter 3, you might also ask what times are best for health promotion (before work? after work? lunchtime?), and whether employees are willing to bear part of the cost. There may be other pertinent questions specific to your organization; be sure to *ask* people for their opinion, rather than assuming you know what they want.

Key decision makers should also have a say at this point. What does your boss or your boss's boss think of the idea? Will they support the program? Do you need to modify or shift the emphasis of the program so that it will fit in with the organization's goals?

Norms might also determine the success or failure of a health-promotion effort. Determine what existing norms and values at your workplace might support or sabotage a wellness program. These might be negative health norms, as discussed earlier, or they might involve management style.

GET THE RIGHT PEOPLE INVOLVED

Health promotion isn't something you do alone. It's an organizational effort, usually involving members of different departments. A typical health-promotion "task force" might include members of the medical and human resources departments, senior management, union representatives, security personnel, employees and others. While it may be difficult to coordinate such a variety of planners, cooperation and support from a wide range of people in the organization is essential.

There's no such thing as a "turnkey" health-promotion program that you can buy complete from a consulting firm. Successful health promotion requires involvement from *within* the organization, preferably from all levels.

GET CLEAR ON YOUR OUTCOME

The ultimate goal of any wellness program is, of course, to help participants become healthier and stay healthier. But there are a number of other outcomes that are important to the success of current and future programs. The following criteria can help you decide where to start. By asking yourself, "How important is each of the following?" you can determine what type of wellness program will help you meet your goals.

- *Measurable outcome.* Do you need facts and figures to prove the success of a program? Smoking cessation, blood pressure screening, and fitness programs usually provide results which can be quantified, while the benefits of a stress management program are more difficult to measure.

219

- *High-impact potential.* Those programs that are likely to have the greatest impact on employee health and organizational productivity are alcohol/drug- or smoking-cessation programs.

- *Number in potential target group.* This depends on the results of your needs assessment survey. What sort of programs will be useful to large numbers of people in the organization? Some programs like nutritional education have a broad application to most employees.

- *Likelihood that people will get involved.* In general, the most popular wellness programs are stress management, weight loss, and fitness.

- *Visibility of program.* How important is it to attract corporate attention for health-promotion efforts? Screening programs for cancer or "health fairs" are two programs that have high visibility.

- *Likelihood of short-term and long-term success.* Traditional weight-loss programs (those without an exercise component) have the worst long-term success rates, with fewer than 5 percent of the participants maintaining their weight loss one year later. Smoking-cessation rates are much better. In a well-designed program, 40 to 50 percent of the participants may remain cigarette-free after six months, and 15 to 30 percent will remain off cigarettes after one year.

BUILD A DATA BASE TO MEASURE RESULTS

A wellness program may be aimed at a number of problems: absenteeism, disability or medical costs, employee morale, or general well-being. Establishing a data base that shows where you stand now will help you measure the outcome of the program.

Data may be obtained from the company's health-care insurer or personnel department. Such information might include:

- The cost of health benefits per employee per year
- Absenteeism by total days per year, by days per employee, and by department
- Number of workers' compensation claims per year
- Number of hospital days per employee and per dependent
- Number of hospital admissions by diagnosis per year

You can also use health risk appraisals to gather data for larger populations in your employee group. Based on age, sex, and lifestyle habits, what risks are most prevalent in your company? A health-promotion program might be aimed at lessening those risks through awareness or lifestyle modifications.

There are several different methods of approaching the data-gathering and evaluation phase. The following four designs are the most commonly used methods. Each will allow you to acquire meaningful information. Any single approach or a combination of the four may be used.

- *Historical, record-keeping.* This method shows what happened in the target group over a specific period. It involves collecting data on participation, attrition, and behavior changes.

- *Stop everything, inventory.* This method consists of a follow-up "snapshot" of behavior changes taken at intervals, such as immediately following the program, and at three, six, and twelve months later. Like the first approach, this focuses only on the target group.

- *Comparative: "How we stack up against others."* This method extends the "historical" approach by comparing data on the target group against results from other health-promotion programs. This helps isolate outside factors and evaluate the success of a specific program.

- *Controlled-comparison, quasi-experimental design.* This approach compares participants in wellness programs against a "control group" not receiving the program. In large organizations, this approach can be extremely valuable in quantifying the benefits of a program.[5]

DEFINE YOUR TARGET AUDIENCE

The details of a health-promotion program will vary considerably according to the target audience. Its size, educational background, job duties, even gender or ethnic makeup will create special needs and opportunities.

A program aimed only at upper-level managers, for instance, might be more comprehensive because of the smaller size of the target group. Such a program could provide more thorough physical examinations and individual counseling than a company-wide effort could. A program for middle- and first-level managers, on the other hand, might focus more on communications skills, or perhaps on alcohol or drug training. A general program aimed at all employees must take into account the special needs of the group. For instance, a work force that is largely female would benefit from training in breast self-examination.

Whatever the audience, a successful health-promotion program responds to the needs and interests of people in the group.

DETERMINE THE SCOPE OF THE PROGRAM

How extensive will the program be? Wellness efforts can vary from informational campaigns to full-scale intervention efforts. Determining which one suits your needs depends on the budget and staff you have available, as well as the level of commitment and enthusiasm in your work force. In general, the programs that provide intervention to improve health or eliminate risk factors (for instance, stop-smoking programs) involve the most effort, but are most effective in getting results.

Awareness efforts are a good place to begin if your organization is just becoming involved in health promotion. George Pfeiffer, immediate past President of the Assocation of Fitness in Business (AFB) and Vice President of the Center for Corporate Health Promotion in Reston, Virginia, sees the communications explosion as a positive force for health promotion: "[I see] no reason why any company, regardless of size, can't have an effective health promotion effort based upon a properly implemented communications based model."[6]

The communications network in your company provides almost unlimited opportunities for spreading the word about health. Notices on bulletin boards, articles in the company newsletter, promotional calendars or handouts, paycheck stuffers, noontime seminars, and videotapes are some ways to create greater awareness of health issues.

The effectiveness of educational efforts unaccompanied by intervention was shown at the Kaiser Permanente Medical Care Program in Portland, Oregon. A recent evaluation of a three-year series of health-education campaigns involving approximately 3,000 employees in eighteen locations showed that low-expense efforts to promote awareness can have an effect on behavior. The Kaiser effort is credited with influencing the behavior of between 8 percent and 15 percent of those who received the materials, as shown in the following table:

PERCENT OF RESPONDENTS REPLYING IN THE AFFIRMATIVE

	Blood Pressure	Nutrition	Smoking	Exercise	Breast Cancer Prevention
1. Did you receive campaign materials?	61.7	69.0	74.5	81.5	83.1

Establishing Wellness Programs in Your Company

	Blood Pressure	Nutri- tion	Smok- ing	Exer- cise	Breast Cancer Preven- tion
2. Did you read campaign materials?	53.3	63.1	46.7	73.9	72.9
3. Did the campaigns prompt you to take action?	13.5	20.6	6.4	32.7	28.9
4. Did the campaigns affect behavior change very much?	9.1	10.1	8.0	16.4	15.1
5. Should the campaigns be continued or expanded?	65.0	66.4	68.9	70.9	69.1

FIND OUT WHAT YOU CAN DO WITH LOCAL RESOURCES

In many cases, wellness programs can be coordinated and staffed solely with internal resources. For more ambitious programs, though, you may need to recruit additional outside help. Before you call in the "experts," determine how extensive their experience is, and how expensive their help will be.

Many organizations can help you set up a low-cost health-awareness effort. Hospitals, YMCA's, heart and lung associations, and health-insurance companies often have free or low-cost materials or programs to disseminate health information. Often, these organizations will donate resources for a "health fair," where information on a variety of health topics will be available.

BUILD SUPPORT GROUPS

When you decide to go beyond awareness efforts into a more active health-promotion program, support and leadership become vital. Peer pressure can be a powerful tool for achieving and maintaining success in health promotion, and a well-led group can help many individuals who wouldn't be able to succeed alone.

The success of the group, though, often depends on the leader. An effective leader must maintain interest, protect confidentiality within the group, provide a high level of acceptance, and preserve a balance of participation among the members. The leader can encourage participation from less aggressive members and guide the discussion to prevent diverting the group from its purpose. Participants should be encouraged to come to meetings *especially* when they are not doing well. Most behavior-change efforts encounter a "low point" after three or four weeks; the group leader should be prepared to keep enthusiasm going through this slump.

Support groups vary in number of members, duration, format, and frequency of meetings. Groups of between eight and twelve members offer several advantages: Meetings are usually interesting and lively, and if the group continues for ten to twelve weeks, each participant will be able to act as a leader. If time for meetings is limited (i.e., a half-hour lunchtime meeting), smaller groups of four to six members are more appropriate.

At each meeting, members share their records of progress and discuss how they are doing. A group may discuss a particular topic or share materials, or group members can take turns leading the group or giving presentations.

MAKE SURE LEADERS ARE QUALIFIED

For programs involving skill enhancement or lifestyle changes (for instance, stop-smoking programs or stress management seminars) many organizations choose to hire consultants who specialize in these programs. When you "buy" a health-promotion program, the quality of the instruction often determines its success. More programs fail because of poor instruction than for any other reason. A good health-promotion instructor should be:

> *A role model*, yet able to empathize with those who haven't yet achieved their goals. A good instructor energizes and inspires class members to change, without being "preachy" or resorting to a military-type bullying attitude. There is such a gap between the fit and the non-fit, between smokers and non-smokers, that the best instructors are those who can empathize with the difficulty of achieving health changes, preferably through having made the changes themselves.

> *Experienced at working with people*, especially those in management. The instructor needs to be able to relate to people on all levels.

227

A *good communicator,* preferably with a sense of humor. Ask to see a videotape of the instructor in action, attend a class, or get references.

In addition to checking out the instructor, review any audio-visual materials or handouts that will be used in the class. Find out what the class objectives are, what the structure of the class is, how they plan to reach people. Ask what the instructor's follow-up plans are for reinforcing the new skills.

PERKS: INDIVIDUAL HEALTH-PROMOTION TOOLS

A well-planned, carefully executed health-promotion program can go a long way toward promoting wellness in the workplace. Yet even the best program is only part of the solution. For maximum well-being and productivity, employees need to be managed for health; that is, they need PERKS.

When it comes to managing for wellness and productivity, PERKS and health promotion go hand-in-glove. Each approach alone has merits, but when the two are combined, each enhances the effectiveness of the other.

There is no single path to wellness at the workplace. PERKS and health promotion can be thought of as a twofold path: When both approaches are used, you can cover more ground and reach your dual goal— wellness and productivity—more quickly and more surely.

NOTES

Chapter 1
1. These statistics are drawn from a number of sources. While not exact, they represent the best estimates based on current research:

   National Health and Nutrition Examination Survey, II, National Center for Health Statistics (Rates Applied to Employee Population), 1981.

   Smoking at the Workplace: A Program Guide (New York: American Lung Association, 1981).

   President's Council on Physical Fitness.

   National Health and Nutrition Examination Survey, 1971–74, Advanced Data, #51, National Center for Health Statistics.

   Lester Breslow, "Risk Factor Intervention for Health Maintenance," *Science*, Vol. 200, May 26, 1978, pp. 908–12.

   American Heart Association Monograph #60, "Relationship of Blood Pressure, Serum Cholesterol, Smoking Habits, Relative Weight and ECG Abnormalities and Incidence of Major Coronary Events."

Cardiovascular Primer for the Workplace, 1981. Health Education Branch, Office of Prevention, Educational Control; National Heart, Lung and Blood Institute, NIH Publication No. 81–2210.

2. *National Heart, Lung and Blood Demonstration Projects in the Workplace: High Blood Pressure Control*, draft paper prepared May 1983.

   Joan Beck, editorial in Chicago *Tribune*, April 23, 1984.

   Statistics from the President's Council on Physical Fitness and Sports.

   Reprinted by permission of *Harvard Business Review*. Excerpt from "Drugs in the Workplace," by Peter Bensinger (November–December 1982), © 1982 by the President and Fellows of Harvard College; all rights reserved.

   T. B. Van Itallie, "Obesity: Adverse Effects on Health and Longevity," *American Journal of Clinical Nutrition*, 32 (1979), pp. 2723–33.

3. E. C. Hammond, "Smoking in Relation to the Death Rates of One Million Men and Women," in *Epidemiologic Approaches to the Study of Cancer and Other Chronic Diseases*, W. Haenszel, ed. Bethesda, Md.: National Cancer Institute, 1966; U.S. Department of Health, Education, and Welfare, 1979a; U.S. Department of Health and Human Services, 1980b.

4. J. C. Erfurt and A. Foote, "Final Report: Hypertension Control in the Worksetting," The University of Michigan, Ford Motor Company Demonstration Project Submitted to the Heart, Lung and Blood Institute, NIH, DHHS, Contract No. NO1–HV–8–2913, 1982.

5. *Training Aids Digest*, Vol 9, No. 4, April 1984. Reprinted with permission.

6. Interview with William Smithburg conducted by Mark Tager, November 1, 1984.

7. Interview with Warren Batts conducted by Mark Tager, November 26, 1984.

Chapter 2

1. According to an article in the *Wall Street Journal* dated September 17, 1980, three to four thousand workers' compensation cases are filed each year in the state of California alone

Notes

for stress-related disabilities. Of these, 50 percent are decided in favor of the claimant.

In extreme cases, manager-caused stress may be fatal. In a June 1981 *Psychology Today* article, Senior Editor Berkeley Rice in "Can Companies Kill?" cites a case where a businessman's widow was suing his former employer, claiming that his suicide was a direct result of stress caused by his superiors at work.

2. G. E. Vaillant, *Adaptation to Life: How the Best and Brightest Came of Age* (Boston: Little, Brown, 1977).

3. George Engel, "Emotional Stress and Sudden Death," *Psychology Today*, November 1977.

4. Rosabeth Moss Kanter, *The Change Masters: Innovation for Productivity in the American Corporation* (New York: Simon & Schuster, 1983), p. 62.

5. J. M. Weiss and H. I. Glazer, "Effects of Acute Exposure to Stressors on Subsequent Avoidance-Escape Behavior." *Psychosomatic Medicines*, 37 (1975), pp. 499–521.

6. Robert M. Yerkes and John D. Dodson, "The Relation of Strength of Stimulus to Rapidity of Habit-Formation," *Journal of Comparative Neurology and Psychology*, 1908, p. 459.

7. D. O. Hebb, *A Textbook of Psychology* (Philadelphia: Saunders, 1958).

8. R. D. Kaplan, *Job Demands and Worker Health: Main Effects and Occupational Differences* (Washington, D.C.: U.S. Department of Health, Education and Welfare, 1975).

9. We can see from the examples below how job structure can cause the physical/chemical reactions of stress:

    Too much to do/too little time: A study of office workers whose jobs involved frequent overtime work found that increased epinephrine levels (another physical reaction to stress) were present both during and after working hours. A. Rissler, "Stress Reactions at Work and After Work During a Period of Quantitative Overload," *Ergonomics*, 20 (1977), pp. 13–16.

    The pressure of deadlines: A classic study showed marked increases in the blood cholesterol levels of tax accountants as the April 15 filing deadline approached. After the deadline, their cholesterol levels gradually returned to normal. M. Friedman, R. Rosenman, and V. Carroll, "Changes in the

Serum Cholesterol and Blood Clotting Time of Men Subject to Cyclic Variation of Occupational Stress," *Circulation,* 17 (1957), pp. 852–61.

Shift work: Those who work evenings or nights report more sleep, mood, digestive, and social problems than their counterparts who work during the day. M. Frankenhaeuser, "Coping with Job Stress—A Psychobiological Approach," in B. Gardell and G. Johansson, eds., *Working Life: A Social Science Contribution to Work Reform* (London: Wiley, 1981).

10. J.R.P. French and R. D. Caplan, "Psycho-Social Factors in Coronary Heart Disease," *Industrial Medicine,* 39 (1970), pp. 383–97.

11. E. B. Palmore, "Physical, Mental and Social Factors in Predicting Longevity," *Gerontologist,* Summer 1969, pp. 103–108.

12. Patricia E. Renwick, Edward E. Lawler et al., "What You Really Want from Your Job," *Psychology Today,* May 1978, p. 53.

Chapter 3

1. The "autocratic" and "democratic" organizational styles are based on the four-tiered management systems described by Rensis Likert in *The Human Organization* (New York: McGraw-Hill Book Company, 1967), pp. 4–10.

2. The health-promotion questionnaire on page 85 was developed by Dr. Richard Bellingham at Possibilities, Incorporated, Basking Ridge, New Jersey.

3. Daniel Yankelovich and John Immerwahr, *Putting the Work Ethic To Work: A Public Agenda Report on Restoring America's Competitive Vitality,* 1983, p. 37. Published by the Public Agenda Foundation, 6 East 39th Street, New York, NY 10016. Reprinted with permission.

4. An excellent review by Kenneth E. Warner and Hillary Murt of the Department of Health Planning of the School of Public Health, University of Michigan, shows that ongoing financial reinforcement (such as Aerobic Challenge) achieves better results than programs that offer lump-sum payments. They cite a program at Speedcall Corporation, where employees who do not smoke on the job receive an extra seven dollars